Development of Scientific Research Projects

This textbook offers high school and college students, as well as their instructors, guidance on a number of practical issues, including identifying potential projects, developing a project, designing experiments, collecting and documenting data, analyzing data, and preparing reports, posters, and oral presentations for scientific research projects. It also serves as a reference handbook for scientists and managers who routinely design and execute experiments.

This book helps understand the following:

- Accessible Foundations of the Scientific Method
- Structured Topic Selection through Guided Brainstorming
- Comprehensive Support for Literature Review and Hypothesis Development
- Simplified yet Rigorous Introduction to Experimental Design and Statistical Analysis
- Practical Guidance for Scientific Communication

Humberto Vega-Mercado currently serves as President and CEO of HMV Food and Pharma Consulting, LLC, providing technical guidance to the food and pharmaceutical industries. Throughout his career, he has held senior leadership roles in Manufacturing Science and Technology (MS&T) and related functions. He holds BS and MS degrees in Chemical Engineering from the University of Puerto Rico and a PhD in Engineering Science from Washington State University, and has authored over 40 peer-reviewed publications. Dr. Vega-Mercado is a seasoned leader with over 37 years of experience in the pharmaceutical and food industries, specializing in technology transfers, process validation, and manufacturing. His expertise spans vaccines, sterile and solid dosage forms, food processing technologies, and advanced platforms such as gene and cell therapies. He has led global efforts supporting the successful transfer, validation, and commercialization of multiple therapies, including cell therapies and vaccines, as well as numerous generic drugs.

Development of Scientific Research Projects

Lifecycle Approach

Humberto Vega-Mercado, PhD

CRC Press
Taylor & Francis Group
Boca Raton London New York

CRC Press is an imprint of the
Taylor & Francis Group, an **informa** business

Designed cover image: Shutterstock

First edition published 2027
by CRC Press
2385 NW Executive Center Drive, Suite 320, Boca Raton FL 33431

and by CRC Press
4 Park Square, Milton Park, Abingdon, Oxon, OX14 4RN

CRC Press is an imprint of Taylor & Francis Group, LLC

ISBN: 9781041379126 (hbk)
ISBN: 9781041379119 (pbk)
ISBN: 9781042010516 (ebk)

DOI: 10.1201/9781042010516

Typeset in Minion
by Deanta Global Publishing Services, Chennai, India

Access the Instructor and Student Resources: \\EUW1FS01\bks_digipub\COMPANION_WEBSITE_HANDOVERS\9781041379126_Support Materials

To Dr. Nancy George and Dr. Michael George, who closely supported the realization of the book.

To my wife Donna, who supports me in my professional adventures.

To our children and grandchildren, who are the center of our lives.

To our families and friends who have helped us reach our dreams!

In memory of Eddie, a close friend and brother, you left too early, rest in peace.

Contents

List of Figures

List of Tables

Preface

THIS BOOK IS DEVELOPED to fulfill the need for a guidance book on the subject of scientific research. Throughout my professional career, I have been engaged in scientific research. On multiple occasions, I have wished to have a reference book explaining the fundamentals to develop and conduct scientific research while providing guidance on topic selection, literature review, defining hypotheses, experiment design, data collection and analysis, and publication of the results, while guiding high school and college students.

This book should be used as a reference in the classroom (for high school to college students and teachers), in laboratories by managers and scientists, or as a personal reference for any professional involved in scientific research. The book is mainly focused on a practical use of scientific research in response to students' or scientists' needs while developing and executing scientific projects.

My sincere thanks to students and teachers who are dedicated to learning and practicing science and math. I hope you enjoy the book and make the best use of its content.

Humberto Vega-Mercado

CHAPTER 1

Scientific Research

1.1 INTRODUCTION

For students with limited or no direct experience with the concept of scientific research, the mention of the term "scientific research" may in itself conjure ideas about an eccentric individual wearing a white coat and carrying a big notebook or computer hidden away in a laboratory for hours, day after day, watching some kind of equipment and waiting for some kind of event to happen after which he or she may scream "I found it, I found it." However, that is not the best example of a scientist or what scientific research is about. In reality, we all do scientific and non-scientific research as part of our daily lives in one way or another. Indeed, the concept of research is used by ordinary people in their daily lives. For example, we engage in research when we try to define which computer works faster, which car is more economical or has better fuel economy, which over-the-counter medicine works faster against a cold or allergy, which avocado is ready for use, how much salt is needed in a soup, why a child is crying, which air conditioner is better, or which detergent is more effective at removing stains. As we ask such commonplace questions and seek answers from the observations we take from our environment, we are generally employing the principles of research.

The process of structuring and conducting data collection and analysis with the goal of reaching a specific conclusion makes us day-to-day researchers. However, conducting scientific research is not always as easy as selecting a computer or a detergent. It should be remembered that not

DOI: 10.1201/9781042010516-1

all research activities can be considered scientific. The key question to be asked is: what is the main characteristic of scientific research? To answer this question, we begin with an example:

At home, your mom wants to know if a new detergent is better than the one that she is currently using. As part of the process to determine if the new detergent is better or not, she washes her clothing with both detergents under the same cleaning conditions, for example: (1) follows the manufacturer's instructions for each detergent; (2) uses the same washing machine; (3) uses the same washing cycle (e.g., Cycle # 3 from the dial that lasts 15 minutes using warm water); (4) washes the same type and amount of clothing (e.g., a few jeans, a few shirts); and (5) conducts multiple runs with the new detergent. The outcome of the washing cycles is then evaluated by considering the brightness of the colors and residues of detergent and soil on the clothes. Based on this information, your mom concludes that the new detergent is better than the old detergent.

This case is a simple example of how we conduct research at home. We establish a question, compile information, conduct some experiments, analyze the data, and reach a conclusion. However, we cannot say this is an example of scientific research. Why? In the case of scientific research, we define, control, and monitor additional factors as part of the experiment, such as specific amounts of water and its temperature, source of water, amount of detergent, specific pieces of clothes, and nature of the fabrics, wash time, nature and amount of soil (called independent variables), and define and monitor specific attributes to measure the effectiveness of the cleaning process, such as soil residues on clothes and a standardized measure of color intensity (called dependent variable). These additional considerations were not present in the "experiment" described above that mom ran at home.

The technological advances of the past 100-plus years have resulted in the development of industrial wonders that benefit our society in many areas (e.g., transportation, communications, industrial machines, and pharmaceuticals, among others). On the other hand, we are now confronted with the need to develop new techniques and technical solutions to problems, some of which we created in our efforts to solve past problems. The new challenges are ones we did not anticipate and have emerged in recent years as a result of previous manipulations of our environment (e.g., increased incidence of diseases, greenhouse effect, water and food supply contamination, deforestation, air pollution). In all cases, scientific research provides the mechanism to find either the solution or alternatives to the problems affecting the world every day. As Leonardo Da Vinci (1452–1519) stated:

> *All sciences are vain and full of errors that are not born of experience, the mother of all Knowledge.*

The goal of this book is to provide the reader (e.g., a high school student, a college student, a lab manager, or a lab scientist) with a practical guide for the development, compilation, analysis, and presentation of experimental work as part of scientific research projects. The information provided in the book should be used as a reference while developing and executing research projects. It is important to recognize the limitations of scientific research as well as legal requirements established by different regulatory and law enforcement organizations (e.g., US-FDA, USDA, Law Enforcement, Military, etc.). Therefore, the reader shall be familiar with regulatory requirements as well as any limitations associated with his or her research interests.

The systematic approach to executing scientific research known as "Scientific Method" provides a mechanism for the compilation of information and data analysis that helps overcome some of the limitations any scientist may face during research activities. This approach is the main difference between scientific research and any other type of research (e.g., general surveys).

1.2 GLOSSARY OF TERMS

The following terms, commonly used by scientists, are used in this chapter.

Terms	**Definition**
Artificial intelligence (AI)	AI is the ability of machines or computers to think, learn, and make decisions like humans.
Conclusion	The conclusion is the outcome of the data analysis and either confirms or rejects the hypothesis or hypotheses.
Data analysis	It is the process for collecting, organizing, and interpreting data to answer questions or make decisions.
Experiment	Specific activities intended to prove or disprove a hypothesis.
Hypothesis	Potential explanation, result, or alternative to a problem
Literature	Compilation of historical information and data

Terms	Definition
Observation	Documented results from experiments. Subjective observations are results based on experience and perception of the scientists. Objective observations are actual measurements of a variable under evaluation.
Problem	Issue or question requiring a solution.
Research	A careful and systematic study is done to discover new information or understand something better.
Scientist	A person who studies the world (e.g., biology, chemistry, physics, astronomy, etc.) carefully and uses experiments, observation, and evidence to discover new knowledge.
Scientific method	A systematic approach provides a mechanism for the compilation of information and data analysis.
Variables	Conditions or parameters that are either used to control a process or equipment (independent variables) or to assess the outcome of a process (dependent variables).

1.3 SCIENTIFIC METHOD

The scientific method has multiple definitions. In our case, we define key basic elements or steps to be followed during the development and execution of scientific research. We will call these elements and steps the Scientific Method:

1.3.1 Problem Definition

The first step in initiating scientific research is to define the problem you want to study. The goal of any scientific research is to find the answer to a problem or situation. Questions involving why, what, how, where, or when a particular situation is observed or caused are the basis for scientific exploration. Examples of those questions follow:

- How are cancer cells created?
- Why does the human body develop cancer cells?
- What triggers the growth of cancerous cells?
- What is the impact of reducing the use of coal or burning oil?
- How are trees and plants processing environmental contamination?
- How are chemicals in food affecting babies?
- When are brain cells affected by drugs?
- What is a magnetic field and how does it affect humans?

- What is the effect of pulsed electric fields (PEF) on microorganisms and bacterial enzymatic activity?
- Where is the end of deep space?

1.3.2 Literature Review

Scientific information is available in multiple forms and sources (e.g., journals, books, the internet or web pages, and other electronic sources). Scientists use existing information compiled by other scientists on a particular subject to better understand the facts around the problem under study. The main goal of the literature review is to identify areas requiring further research and experimental work and therefore increase the knowledge available to future researchers. All the information compiled during the literature review is then used to define the hypothesis to be studied by the scientist. Scientific articles that may be considered for one of the problems previously identified (e.g., the effect of pulsed electric field (PEF) on bacteria and bacterial enzymatic activity) are:

- Engineering design of PEF equipment.
- Impact of PEF on protein structures.
- Impact of PEF on microorganisms.
- Conductivity requirements when using PEF.
- Difference between thermal and non-thermal preservation methods.

In summary, the literature review allows one to gain additional knowledge on a subject and narrow the scope of the problem that is to be researched.

1.3.3 Hypothesis

The hypothesis is used to define the potential explanation, result, or alternative to the problem under study. In the case of the example previously discussed (e.g., use of PEF), the hypothesis was:

- Microorganisms and bacterial enzymes can be inactivated using PEF.

Once the hypothesis is postulated, you will develop a plan to conduct experimental work that will either prove it or disprove it. In any case, the scientific research will contribute additional data to be used by other scientists working in the same field. Scientific research is not a race to win or lose; it is a quest for increasing knowledge in the field. In summary, the hypothesis will be proved or disproved based on the data collected and its analysis.

1.3.4 Experiments

Experiments are specific activities intended to prove or disprove the hypothesis. The experimental work is conducted under controlled conditions and detailed in a test protocol. The protocol defines how the different variables are defined, controlled, measured, and evaluated, as well as the steps for conducting the experiments. There are two types of variables in any experiment:

- Independent Variables: Those parameters that can be controlled in an experiment (e.g., time, temperature, pressure, quantities, speed, flow).
- Dependent Variables: Those parameters that are affected by the independent variables (e.g., yield, growth, distance, hardness, color, survival, residues).

In the example we used previously involving PEF for the inactivation of microorganisms and enzymes, the variables used for the experiments were:

- Independent Variables: PEF Intensity (Volts/cm), Frequency of pulses (Hz), Treatment Duration or residence time (seconds; as a function of flow rate and volume of treatment chamber), Temperature of solution (°C), type of media where bacteria and enzymes were suspended (conductivity, concentration of specific compounds), type of microorganisms (e.g., vegetative cells, spores, rods vs. cocci; bacteria, molds, and yeast) and type of enzymes (e.g., intracellular vs. extracellular).
- Dependent Variables: Microbial concentration post treatment (bioburden, CFUs/mL), enzymatic activity post treatment (enzymatic activity measures), organoleptic changes of the media where the microorganisms and enzymes were suspended (e.g., color, appearance, odor, taste).

In summary, the experiment is where additional knowledge is gained on how the different independent variables affect (or do not affect) the dependent variables being measured and evaluated.

1.3.5 Observations

The scientist is looking for the outcomes of the experiments (e.g., response of dependent variables as a function of the independent variables). Scientists use all of their senses, as well as technical resources (e.g., instruments) available to them, to record the responses during the experiment. The process of collecting data is known as "Observation." There are two types of observations:

- Subjective Observations: Results based on the knowledge and perception of the scientist (e.g., color intensity).
- Objective Observations: Results based on actual measurements of the variable under evaluation by means of instruments (e.g., distance, pH, weight, concentration).

In our example of the PEF study, most of the observations were "objective" as specific instruments and techniques were used to measure the different variables. However, subjective observations were also used (e.g., appearance, odor, and taste) by means of a trained group of scientists using carefully defined conditions commonly used in these cases (e.g., hedonic scales).

In summary, observations are the raw data compiled during experiments that will be used to assess the hypothesis.

1.3.6 Analysis

The interpretation of the raw data gained during the experiments is known as "analysis." Analysis is the most important part of the scientific research process, as it will be used to either confirm (prove) the hypothesis or reject (disprove) it.

In the case of the PEF experiments, the results confirmed the capability of pulsed electric fields for the inactivation of microorganisms and enzymes. Also, the experimental work provided evidence that the effect of PEF on microorganisms is different based on their type (e.g., bacterial vs. molds) as well as the nature of the media where the microorganisms and enzymes were suspended.

1.3.7 Conclusion

Once all the data are analyzed, the scientist is in a position to draw a conclusion or conclusions based on the data that were collected. The conclusion either confirms or rejects the hypothesis. Whether the hypothesis is confirmed or rejected, the scientist has increased the knowledge in the subject under study.

1.4 SCIENTIFIC RESEARCH AND SCIENCE

The advance of science is due to the hard work of scientists in different fields (e.g., biology, chemistry, physics, botany, psychology, sociology, etc.) throughout time. The medical field, for example, takes advantage of scientific breakthroughs for the benefit of the human race by developing alternatives for a better life. These breakthroughs occur over time, not in the blink of an eye, taking in some cases years of intense research. Each day is a challenge and hope!

As scientists, we will have both successes and failures. Science is a constantly evolving field. New discoveries, new techniques to conduct analyses, and new technological advances (e.g., equipment, computers) bring the different scientific fields closer. Chemistry, biology, and physics are now combined to describe different phenomena that in the past were explained with multiple limitations by using the knowledge from only one of the fields.

Some examples of how the scientific advances in multiple fields result in a better standard of life are: (1) explaining the relation between behavioral conditions to specific chemical imbalances in the brain or body; (2) the impact of diabetes on overall health; (3) in-depth understanding of microbial contamination and food safety; (4) improved food product shelf life; (5) reduced use or elimination of preservatives from the food chain.

1.5 ETHICS

The ethical aspects of scientific research are not limited to executing experimental work under the existing laws and regulations. The development and publication of scientific research are based on facts and data and should not be manipulated to reflect results that are not correct or misleading. All data shall be reported as compiled. The conclusions shall be based on the data collected during the experimental work and literature review. If additional experiments are required, the scientist shall acknowledge the need for further experiments and proceed accordingly. Keep in mind that other scientists may use published reports to replicate prior experiments, and in doing so, they will either confirm the results or challenge the data and findings of previous research.

1.6 FINAL COMMENTS

Scientific research is the cornerstone of the technological advances known to the human race. Researchers are responsible for understanding the limitations of the research work they are conducting as well as following the ethical aspects associated with the development and execution of scientific studies. Today's dreams are the reality of the future when responsible scientific work is conducted for the sake of a better society. The key to scientific research is to provide humanity with logical explanations and alternatives to the events and phenomena affecting the world every day.

Artificial intelligence (AI) applications (e.g., ChatGPT®, Claude, MS Copilot®, Google Gemini, Notion AI, etc.) are increasingly assisting scientists and research groups in compiling research information as well as analyzing data and formulating conclusions. As with any technological tool, effective use of AI requires appropriate data governance and careful contextual interpretation to ensure reliable and valid results.

CHAPTER 2

Problem Definition

2.1 INTRODUCTION

The goal of any scientific research is to find the answer to a problem or situation. Questions involving why, what, how, where, or when a particular situation is observed or caused are the basis for scientific exploration, as previously mentioned. Therefore, the first step to develop a research project is defining the problem to be studied. **Table 2.1** summarizes some examples of problems or research areas that are widely known.

In several instances, a particular problem may be evaluated from multiple disciplines. Therefore, it is important to properly define the scope of the problem and the goal of the project.

2.2 GLOSSARY OF TERMS

The following terms, commonly used by scientists, are used in this chapter.

Terms	**Definition**
Articles (publications)	Documents published by scientists to communicate their experiments and results.
Biology	Scientific study of living things.
Biotechnology	The use of living organisms or biological systems to make useful products.
Botany	Scientific study of plants.
Brainstorming	Technique used to identify potential topics or areas for research.

DOI: 10.1201/9781042010516-2

Terms	Definition
Chemistry	Scientific study of matter and the changes it undergoes.
Engineering	The use of science and mathematics to design and build machines, structures, and systems
Medicine	The science and practice of diagnosing, treating, and preventing diseases.
Microbiology	Scientific study of microorganisms.
Physics	Scientific study of matter, energy, force, and motion.
Psychology	Scientific study of the mind and behavior.

TABLE 2.1 Examples of Problems across Different Disciplines or Fields

Discipline/Field	Problem
Microbiology	Mutation of viruses Vaccines Microorganism resistance to antimicrobial drugs (superbugs) Spread of diseases within communities
Psychology	Chemical imbalance and behavior in children and adults Root cause of Autism Depression Drug and alcohol addiction
Biology/Botany	Impact of radiation on cells Effect of environmental contamination on crops Insects and diseases Reforestation
Chemistry	Biodegradable plastics and other compounds Interaction of foods with plastics PFAS – for ever plastics
Physics	Sub-atomic particles Bending of light The atomic age – managing atomic waste
Engineering	Solar and Wind power – Replacing fossil fuels Greenhouse effect – Reducing CO_2 production Waste management
Medicine/ Biotechnology	New treatment for cancer Patient customized treatments Chimeric Antigen Receptors – T-Cells

2.3 RESEARCHER'S LIMITATION

While science has no limits, scientists are limited by resources (e.g., money, time, equipment, space, facilities) and regulations. Scientists should remember that their contributions to the scientific field are not measured by how much work is accomplished, but by the quality of the work conducted on behalf of the scientific community. Scientists must keep a balance between what is desired, what is possible, available resources, the complexity of the research, and any controversial issues associated with the topic. The laws and regulations must be fully followed by the scientists. Under no circumstances shall a scientist consider himself or herself above the law. Also, all ethical principles defined by professional organizations shall be followed to avoid any conflicts that could arise during the research activity.

In those cases where there is the potential for ethical or legal issues, the scientist shall take the appropriate steps to communicate, assess, and receive required approvals (e.g., getting government or institution permits or waivers) prior to initiating any experimental activity.

2.4 THE BRAINSTORMING PROCESS

Brainstorming is a technique commonly used to identify potential topics or areas for research. This technique can be summarized in six basic steps:

- Compilation of scientific articles involving the research topic of interest

  This step is intended to identify potential problems requiring further study while getting technical information. Typically, the conclusion of research articles includes potential areas for additional work. Scientists shall be aware of new potential areas of research from existing scientific work.

- Team discussion of the information

  One person's point of view does not represent the overall reality of the subject. It is recommended to have multiple points of view on a subject to better define the scope of the project. Feedback from peers (e.g., students, professors, friends, and other scientists) will provide both positive and constructive points of view that help in defining the project. The feedback must be provided with a scientific background to be valid while justifying or rejecting any idea or topic.

- Narrowing the scope of potential research topics

  Once the ideas have been evaluated and discussed, the scientist shall focus on selecting the ideas that are viable and realistic. Also, the scientist shall assess the resources required to proceed with the experimental work and identify any constraints prior to initiating the project. If the resources are not available, the scientist shall assess the potential for additional help from other institutions or laboratories before removing it from the pool of potential projects. Also, sometimes preliminary laboratory work is required to compile additional information not readily available from the literature. The scientist shall coordinate such work and execute any lab work to justify the project.

- Topic-specific discussions

  At this stage, the goal is to identify additional sources of information and resources to execute the projects. Once the information is compiled, the final project is selected. The selected project becomes the goal, and all efforts are now focused on using the scientific method. The scientist shall feel satisfied and happy with the selection and shall be committed to work on the project.

- Additional data collection (literature) on the topic (Step Two of the scientific method) requires an in-depth literature review on the subject to be investigated. Details about the literature review will be provided in Chapter 3, "Literature Review."

- Identify areas of conflict

  Areas of conflict that will affect a research project are not always obvious. Sometimes potential issues are identified early during the literature review process (e.g., collection of samples, preparation of prototypes, specialized equipment or chemicals, source of microorganisms), while others are faced during the experimental steps (e.g., availability of equipment, broken equipment, complexity of testing procedures). You shall be prepared to face these conflicts and look for alternatives or solutions to the situation.

2.5 DEFINING THE HYPOTHESIS

At this point, you have identified the problem and have compiled enough scientific information (from literature review or preliminary laboratory experiments). The scientist shall be able to define the hypothesis. It shall be clear and simple, presenting a tentative solution to the problem under study. The hypothesis shall present the presence or absence of a particular effect (i.e., null hypothesis H_0 and the alternate hypothesis H_a) as well as the definition of the effect: $\mu_1 = \mu_2$ or $\mu_1 > \mu_2$ (Steel and Torrie, 1980; Montgomery, 1984; Ott, 1993; Kiemele et al., 2000; Pyzdek, 2003).

The following example summarizes the process of defining the hypothesis.

The inactivation of enzymes using pulsed electric fields in milk products is being evaluated. The problem statement is *Pulsed Electric Fields inactivate those enzymes contained in milk products (e.g., proteolytic enzymes from Pseudomonas spp.)*. The hypothesis expressed in terms of H_0 and H_a follows:

$$\mathrm{H_0} : \frac{C}{C_0} = 0$$

$$\mathrm{H_a} : \frac{C}{C_0} \neq 0$$

Where C is the activity of the enzyme after exposing the solution to a particular PEF treatment, C_0 is the activity prior to the treatment.

Once all the experimental work is completed and all data is analyzed, the scientist shall be able to either confirm H_0 or refute H_0 and accept the alternative hypothesis H_a.

2.6 FINAL COMMENTS

The identification of the problem, literature review, and definition of the hypothesis will set the stage for the research project. The scientist and the research team must be committed to proceeding with the project. The selection of the project shall consider common sense as well as being open to potential new ideas and options. In some instances, preliminary laboratory work may be required when there is no significant amount of research work on the subject. In other cases, such preliminary work is not required as enough data is available elsewhere (e.g., publications from multiple research groups). The key to the literature review and the depth of such review is to ensure a clear understanding of the laboratory techniques required to execute the experiments. Finally, the hypothesis shall be clear and simple, presenting a tentative response to the problem under study.

2.7 REFERENCES

Montgomery, D.C. 1984. *Design and Analysis of Experiments.* 2nd Edition. John Wiley & Sons, Inc., New York, NY.

Kiemele, M.J., Schmidt, S.R., and Berdine, R.J. 2000. *Basic Statistics: Tools for Continuous Improvement.* 4th Edition. Air Academic Press, LLC, Colorado Spring, CO.

Ott, L. 1993. An Introduction to Statistical Methods and Data Analysis. 4th Edition. Duxbury Press, Belmont, CA.

Pyzdek, T. 2003. *The Six Sigma Handbook: Revised and Expanded: A complete Guide for Green Belts, Black Belts, and Managers at All Levels.* McGraw-Hill, New York, NY.

Steel, R.G.D., and Torrie, J.H. 1980. *Principles and Procedures of Statistics. A Biometrical Approach.* 2nd Edition. McGraw Hill, Inc., New York, NY.

CHAPTER 3

Literature Review

3.1 INTRODUCTION

After a research topic is selected, the next step is the compilation of existing information from existing literature, otherwise known as a Literature Review. The information to be gathered can be found in published books, professional journals, magazines, and topic-specific web pages. Francis Bacon (1561–1626), an English philosopher, statesman, and pioneer in research, pointed to the relevance of gathering knowledge through multiple quotes in his writings. The following are excerpts from those quotes (www.zaquotes.com, 2026; The Editors of Encyclopaedia Britannica. 2026.):

- "It's not what we eat but what we digest that makes us strong; not what we gain but what we save that makes us rich; not what we read but what we remember that makes us learned; and not what we profess but what we practice that gives us integrity."
- "Who questions much, shall learn much, and retain much."
- "Read not to contradict and confute, nor to believe and take for granted…but to weigh and consider."

These quotes remain as valid today as they were back in the 1600s. The key message is that reading the work of others provides the scientist with a better perception of the next steps in his own research. They also illustrate the importance of previous successes and failures and provide

DOI: 10.1201/9781042010516-3

details about techniques and pieces of equipment that could help in future research activities.

The American inventor Thomas Alva Edison (1847–1931) highlighted the relevance of a literature review when he stated, "First, study the present construction. Second, ask for all past experiences…study and read everything you can on the subject" (www.zaquotes.com, 2026).

A literature review consists of a careful and in-depth examination of publications on the topic under consideration. The main goal is to assess the level of knowledge and understanding of previous research on the topic. This review will allow the researcher to better frame the scope and purpose of the proposed research activities, including the development of the problem statement or hypothesis. Specifically, the review allows the scientist to: (1) identify previously completed research studies; (2) provide clear context for the proposed research; (3) identify any limitations, constraints, or challenges already identified in previous work; (4) define the variables to be considered in the study; (5) define the methods and equipment to be used; (6) define approaches for data analysis; (7) outline the reports and publications resulting from the research.

Once the information is compiled, a summary of the literature can be created to consolidate the learnings obtained from reading and analyzing the information. It is common to break down the review either in a chronological form or by specific elements (e.g., techniques, methods, results, conclusions) as suggested by Gay (1992).

In this chapter, we provide guidance about the literature review and the compilation of relevant information.

3.2 GLOSSARY OF TERMS

The following terms, commonly used by scientists, are used in this chapter.

Terms	Definition
Bibliography	A list of books, articles, and other sources used in writing a project or research.
Database	An organized collection of data that can be stored, managed, and accessed easily.
Journal	A publication that contains research articles written by experts in a particular field.
Record	Written or stored information about a person, event, or activity.

Terms	Definition
Spreadsheet	A computer program used to organize and calculate data in rows and columns.
Textbook	A book used for studying a particular subject in school or college.
Word Processor	A computer program used to create, edit, and format text documents.

3.3 COMPILING AND REVIEWING LITERATURE

The information obtained to support a research project should be organized considering the relevance to the topic under evaluation (e.g., age, methods, results, conclusions). It is highly recommended to gather a copy of the articles (e.g., paper or electronic, while protecting the copyrights of the authors) with complete bibliographic details (e.g., author(s), year, title, name of journal, book, or magazine, and publication location – city and country). This information is important when creating a list of references supporting the proposed research activities. The advancement in computer technology allows scientists to create and save databases or records using software containing key information from each bibliographical reference. **Tables 3.1, 3.2, and 3.3** provide a list of potential software available for data processing (Word Processors, Spreadsheets, and Databases).

Artificial intelligence (AI) applications (e.g., ChatGPT®, Claude, MS Copilot®, Google Gemini, Notion AI, etc.) are increasingly assisting scientists and research groups in compiling research information as well as analyzing data and formulating conclusions. Every AI user needs to "train" the AI application to his/her style as well as the type of information being requested. Nevertheless, as with any technological tool, effective use of AI requires appropriate data governance and careful contextual interpretation to ensure reliable and valid results. Each AI is optimized differently. Some prioritize safety constraints, some prioritize speed, some prioritize enterprise integration, and some prioritize personality. **Table 3.4** provides a summary of AI chatbots and assistants available in the market and their typical functionality.

The format of the bibliographic record is, in general, standardized with minor changes considering the publisher or scientific organization. **Table 3.5** provides a summary of formats commonly utilized for your reference.

TABLE 3.1 Example of Word Processors Available for Data Storage

Word Processor	Key Capabilities	Supplier/Developer
Microsoft Word	Advanced formatting, templates, collaboration, cloud sync, AI-assisted editing	Microsoft (part of Microsoft 365)
Google Docs	Real-time collaboration, auto-save to cloud, cross-device access	Google (part of Google Workspace)
LibreOffice Writer	Full offline editing, open-source, wide file format support (DOCX, ODT, PDF export)	The Document Foundation
Apple Pages	Sleek design tools, templates, iCloud sync, strong multimedia integration	Apple
WPS Writer	Word-like interface, excellent MS Office compatibility, PDF tools	Kingsoft (WPS Office)
Zoho Writer	Cloud-based editing with business integrations, automation tools	Zoho Corporation
Scrivener	Organizing long-form writing (books, research), project management	Literature & Latte
OnlyOffice Document Editor	Strong collaborative editing, enterprise features, open source + commercial editions	Ascensio System SIA
FocusWriter	Distraction-free writing environment for focused drafting	(Various community developers)
AbiWord	Lightweight, fast basic word processing	(Open-source community)

3.4 SOURCE OF SCIENTIFIC INFORMATION

The number of sources for scientific information is significant. These sources include journals, books, online publications, magazines, newspapers, and libraries. However, not every potential source of information provides the scientist with the specific information needed for the research activity. With the advances in digital technologies (e.g., Augmented Intelligence or Artificial Intelligence) and online repositories for scientific investigations, the search for published information is easily available. **Tables 3.6 and 3.7** list few publications (journals and books) in different scientific fields.

TABLE 3.2 Example of Spreadsheets Available for Data Storage and Processing

Spreadsheet Program	Key Capabilities	Supplier/Developer
Microsoft Excel	Advanced formulas, pivot tables, charts, data analysis, macros/VBA, strong ecosystem	Microsoft
Google Sheets	Real-time collaboration, cloud-native, cross-platform access, basic to moderate analytics	Google LLC
LibreOffice Calc	Open-source, supports pivot tables, macros, extensive Excel file compatibility	The Document Foundation
Apple Numbers	Visually rich spreadsheets with flexible layout, intuitive interface (macOS/iOS)	Apple
Zoho Sheet	Cloud-based, real-time collaboration, integrations, automation tools	Zoho Corporation
Smartsheet	Spreadsheet + project management, Gantt charts, workflows/automation	Smartsheet Inc.
Airtable	Hybrid spreadsheet-database, multiple views (grid, calendar, kanban), flexible fields	Airtable (now part of the Salesforce ecosystem)
ONLYOFFICE Spreadsheet Editor	Supports Excel formats, pivot tables, data validation, integrated with the workspace	Ascensio System SIA (ONLYOFFICE)
EtherCalc	Simple, real-time collaborative spreadsheet in browser, no sign-in needed	Open-source project
PlanMaker	Strong Excel-compatible functions, .xlsx support, pivot tables, scripting	SoftMaker

Another important source of information includes master's and doctoral theses and dissertations by graduate students. These works are found at local university libraries and via online services such as ProQuest (http://about.proquest.com/en/dissertations/). Dissertations are focused on specific topics with significant and relevant literature reviews.

TABLE 3.3 Example of Databases Available for Data Storage

Database Software	Supplier/Vendor	Key Capabilities
Oracle Database	Oracle Corporation	Enterprise-scale OLTP/OLAP, strong security, in-memory analytics, multi-model support
MySQL	Oracle (Stewardship), originally MySQL AB	ACID transactions, easy setup, wide ecosystem, strong community
Microsoft SQL Server	Microsoft	Strong BI integration, enterprise transactions, analytics, hybrid cloud
PostgreSQL	PostgreSQL Global Development Group/EDB (enterprise)	Advanced SQL, extensibility, complex queries, ACID compliance
MongoDB	MongoDB, Inc.	Flexible schema, JSON-like document store, scalable sharding
Redis	Redis Ltd (original open-source)	Extremely fast in-memory key/value store, caching, real-time
Snowflake	Snowflake Inc.	Cloud data warehousing, scalable analytics, separation of compute/storage
IBM Db2	IBM	Enterprise workloads, high availability, analytics, AI integration
Elasticsearch	Elastic NV	Distributed search & analytics, real-time query performance
SQLite	Public domain (D. Richard Hipp)	Lightweight, zero-config, embedded file-based SQL engine

3.5 FINAL COMMENTS

The literature review process is critical to gain an in-depth knowledge and understanding of the research topic under consideration. A thorough and careful review and data compilation will help define the scope and definition of the hypothesis to be evaluated.

Copyright laws and plagiarism issues must be kept in mind when using information from other scientists. Specifically, any information used from published materials must be properly attributed to the original author. This same consideration applies to diagrams and pictures.

TABLE 3.4 Examples of AI Tools and Their Functionality

AI Application	Main Functionality
ChatGPT	A large language model-based AI system designed to understand and generate human language, and to assist with reasoning, content creation, analysis, and problem-solving across domains
Claude (Anthropic)	A conversational AI focused on *safety, controlled responses, and long-form reasoning*; often used for writing and analysis tasks
Google Gemini	Google's versatile AI assistant handles text, images, and multimodal interaction, tightly tied into Google services.
Microsoft Copilot / Bing AI	Integrated deeply into *Microsoft products* (Word, Excel, Outlook) and search, useful for productivity and facts.
Perplexity AI	Chat + real-time web search, showing sourced answers useful for research and accuracy.
DeepSeek	A newer generative AI chatbot with strong performance in reasoning and a lightweight infrastructure.
Ernie Bot (Baidu)	A popular AI chatbot in China based on Baidu's ERNIE models.
GigaChat	A Russian chatbot developed by Sberbank with multimodal capabilities.
Amazon Q	Amazon's enterprise-focused AI assistant for business workflows.
Meta AI	Meta's conversational AI across Facebook/Instagram applications.
Jasper AI	Tailored for marketers, optimized for *copywriting and creative content.*
Character.AI	AI chatbots with *personality-driven interactions* for fun or entertainment.
YouChat (You.com)	A search + AI chat tool combining search engine results with conversational responses.
Notion AI	AI inside Notion for notes, summaries, and creative writing.
GitHub Copilot/Codeium/Amazon CodeWhisperer	AI coding assistants for developers.
Voice & Device Assistants	Siri, Google Assistant, and Alexa (plus newer versions like Alexa+) for voice-based tasks, devices, and routines.

TABLE 3.5 Example of Formats for Bibliographic Records

Publication/Link	Format	Comments
Journal of the Franklin Institute	Guidance: http://www.sciencedirect.com/journal/journal-of-the-franklin-institute/publish/guide-for-authors Article: Author(s). Title. Journal. Volume, (year). Web link Book: Author(s). Book Title. Edition, Publisher. Location. Year Chapter in Book: Author(s). Chapter Title. "In" Book's editor(s), Book Title, Publisher, locations, year, pages. Web Page: Name of the Organization, complete the www link to the information.	References are numbered.
Journal of Food Science and Food Technology Magazine	Use the Chicago Manual of Style: https://www.chicagomanualofstyle.org/ tools_citationguide/citation-guide-2.html Article: Author(S). Year. "Title." Journal volumes (no): pages. http link for publication. Book: Author(s). Year. *Title*. Publisher. Chapter of book: Author(s). Year. "*Chapter Title*." In *Book Title*, edited by Editor(s). Publisher.	References listed in alphabetical order.
Parenteral Drug Association (PDA)	Use the American Chemical Society Style Guide: https://pubs.acs.org/doi/full/10.1021/acsguide.40303 Article (Journal or Magazine): Author(s). Title. Journal name, year, volume number, and page number(s) Book: Author(s). Title of the book, volume, edition, name of publisher, place of publication, year of publication and, if relevant, page number(s)	Articles (journals & magazines) and books are listed first. Other sources follow in alphabetical order.

TABLE 3.6 Example of Publications: Scientific Journals

Journal Name	Publisher	Primary Scope
Nature	Springer Nature	Multidisciplinary science
Science	AAAS	Multidisciplinary science
Cell	Elsevier	Life sciences and biology
The Lancet	Elsevier	Medicine and public health
New England Journal of Medicine (NEJM)	Massachusetts Medical Society	Clinical medicine
Nature Medicine	Springer Nature	Translational and clinical medicine
Proceedings of the National Academy of Sciences (PNAS)	National Academy of Sciences	Multidisciplinary science
Nature Reviews Molecular Cell Biology	Springer Nature	Molecular and cell biology (reviews)
JAMA (Journal of the American Medical Association)	American Medical Association	Medicine and health policy
Nature Biotechnology	Springer Nature	Biotechnology and applied life sciences
Journal of Food Science	Institute of Food Technology	Food science
PDA Journal	Parenteral Drug Association	Pharmaceuticals

TABLE 3.7 Example of Publications: Text and Specialized Books

Book Name	Publisher	Primary Scope
Handbook of Food Preservation (M.S. Rahman, 1999)	Marcel Dekker	Food engineering
Food Chemistry (O. Fenneman, 1985)	Marcel Dekker	Food chemistry
Principles of Food Processing (D.R. Heldman, R.W. Hartel, 1997)	Chapman & Hall	Food engineering
Dehydration of Foods (G.V. Barbosa and H. Vega, 1996)	Chapman & Hall	Food engineering
Enzyme Assays for Food Scientists (C. E. Staufer, 1989)	Van Nostrand Reinhold	Food sciences
Principles of Food Chemistry (J.M. deMan, 1990)	Van Nostrand Reinhold	Food chemistry
Technical Report No. 26: Sterilizing Filtration of Liquids (PDA, 2025)	Parenteral Drug Association	Pharmaceuticals
Filtration in the Biopharmaceutical Industry (T.H. Meltzer and M.W. Jornitz, 1998)	Marcel Dekker	Pharmaceuticals
Guide: Advanced Therapy Medicinal Products (ATMPs) – Autologous Cell Therapies (ISPE, 2021)	International Society for Pharmaceutical Engineering	Pharmaceuticals – cell therapies
Handbook of Cell and Gene Therapy (H. Aranha and H. Vega, 2023)	CRC Press. (Taylor & Francis Group)	Cell and gene therapies
Pharmaceutical Filtration. The Management of Organism Removal (T.H. Meltzer and M. W. Jornitz, 2006)	Davis Healthcare International Publishing	Pharmaceuticals
Principles of Parenteral Solution Validation. A Practical Lifecycle Approach (I. Gorsky and H. S. Baseman, 2020)	Academic Press	Pharmaceuticals

3.6 REFERENCES

Aranha, H., and Vega-Mercado, H. 2023. *Handbook of Cell and Gene Therapy.* Taylor & Francis Group, New York.

Bacon, F. 2026. Francis Bacon Quotes About Science. *A-Z Quotes.* https://www.azquotes.com/.

Barbosa-Canovas, G.V., and Vega-Mercado, H. 1996. *Dehydration of Foods.* Chapman & Hall, New York.

deMan, J.M. 1990. *Principles of Food Chemistry.* Van Nostrand Reinhold, New York.

Edison, T.A. 2026. Thomas A. Edison Quotes About Science. *A-Z Quotes.* https://www.azquotes.com/quotes/topics/science.html.

The Editors of Encyclopaedia Britannica. 2026. Francis Bacon – Thought and Writings. *Encyclopaedia Britannica*. https://www.britannica.com/biography/Francis-Bacon/Thought-and-writings.

Fenneman, O. 1985. *Food Chemistry*. Marcel Dekker, New York.

Gay, L.R. 1992. *Educational Research: Competencies for Analysis and Application*. 4th Edition. MacMillan Publishing Company, New York.

Gorsky, I., and Baseman, H.S. 2020. *Principles of Parenteral Solution Validation: A Practical Lifecycle Approach*. Academic Press, New York.

Heldman, D.R., and Hartel, R.W. 1997. *Principles of Food Processing*. Chapman & Hall, New York.

ISPE. 2021. *Guide: Advance Therapy Medicinal Products (ATMPs) – Autologous Cell Therapies*. ISPE, Tampa Florida.

Meltzer, T.H., and Jornitz, M.W. 1998. *Filtration in the Biopharmaceutical Industry*. Marcel Dekker, New York.

Meltzer, T.H., and Jornitz, M.W. 2006. *Pharmaceutical Filtration: The Management of Organism Removal*. Davis Healthcare International Publishing, New York.

PDA. 2025. *Technical Report No. 26: Sterilizing Filtration of Liquids*. PDA, Bethesda.

Rahman, M.S. 1999. *Handbook of Food Preservation*. Marcel Dekker, New York.

Staufer, C.E. 1989. *Enzyme Assays for Food Scientists*. Van Nostrand Reinhold, New York.

CHAPTER 4

Design of Experiments

4.1 INTRODUCTION

Scientific research requires proper design, planning, and execution of experiments. The planning and design phases include the following elements:

- Problem Statement: Scope and purpose are well understood, clear hypothesis.
- Identification of the Variables: Justification of the independent and dependent variables. A proper control strategy for independent variables shall be in place to ensure proper data analysis once data is collected (e.g., dependent variables).
- Nature of Experiments: Aligned with the hypothesis.
- Sampling Plans: Defining the type and number of samples to be collected and tested based on sound statistical principles.
- Data Analysis: Proper use of statistical tools aligned with the experiment, sampling plans, and hypothesis.
- Conclusions: The result of the experimental work is assessed considering the outcome of the data analysis.

The goal of this chapter is to provide guidance on the design of experiments using general principles commonly found in science and engineering fields.

DOI: 10.1201/9781042010516-4

4.2 GLOSSARY OF TERMS

The following terms, commonly used by scientists, are used in this chapter.

Terms	Definition
Analytical methods	It encompasses both classical techniques (e.g., titration, gravimetric analysis) and modern instrumental approaches (e.g., spectroscopy, chromatography, mass spectrometry, and electrochemical methods).
CD4+ and CD8+ T-cells	Specific human T-cells (type of leukocytes) in the blood stream responsible for attacking and destroying extraneous particles, microorganisms, and malignant cells as part of an immune response.
Colony forming units	Clusters of microorganisms observed on culture media.
Culture media	Special solutions and formulations used to grow and maintain living cells and microorganisms.
Design of experiments	The design of any task that aims to describe and explain the variation of information under conditions that are presented in a hypothesis. Examples include *Completely Randomized, Randomized Block, Latin Squares,* and *Factorial.*
Distribution	Particular value or value range of a variable.
Empirical rule	Used to define the shape of a normal distribution of data with a defined mean and standard deviation values.
Environment	The surroundings of a physical system that may interact with the system by exchanging mass, energy, or other properties.
Freezing	Cooling of products or materials below 0°C, commonly known as the freezing point of water.
Homogenization	Process utilized to create a uniform distribution of particle sizes in a solution or suspension.
Instruments	Measuring tools used in a scientific laboratory.
Laboratory	Place or facility where experiments can be conducted.
Mean	Average of the measures.
Median	The central point 50% of measures are above and 50% are below this value.
Mode	Most frequent value from a data set.
Morphology of cells	Shape and physical characteristics of cells and microorganisms.
Non-thermal treatment	Process intended for the inactivation of microorganisms without a significant increase in product temperature. Examples of these processes include pulsed electric fields, pulsed magnetic fields, high pressure, preservation using hurdle technologies.

Terms	Definition
Organoleptic	Characteristics such as odor, color, and flavor of a material.
Viability/viable	Cells with active metabolic functions and able to multiply per their lifecycle.
Viable microorganisms	Microorganisms with active metabolic functions and able to multiply through their lifecycle.
Pasteurization	Heating process to inactivate vegetative microorganisms in liquids (e.g., foods) using a temperature of 80–90°C for a limited amount of time. It will not kill all microorganisms or spores.
Percentile	The nth percentile is defined as the value in which a percentage of the data is below the such value.
Population	It is the whole group of items or elements of interest.
Problem statement	Concise description of an issue or lack of knowledge to be investigated.
Random	Lack of a definite pattern or predictability in information.
Range	Difference between extreme values observed in a data set.
Red blood cells (RBC)	Type of cells in the bloodstream. Red color associated with the presence of hemoglobin.
Safety	The state of being protected from harm or other danger. Safety can also refer to the control of known hazards in order to achieve an acceptable level of risk.
Sample	A subset of a population of interest.
Sampling	Act of collecting samples.
Shelf Life	Total number of minutes, hours, days, months, or years a material may last before it is no longer useful.
Standard deviation	The positive square root of the variance.
Statistics	A discipline that concerns the collection, organization, analysis, interpretation, and presentation of data.
Variables: discrete, continuous, qualitative, and quantitative	Discrete: Variable with a finite number of values Continuous: Variable with an infinite number of values. Qualitative: Variable that can vary in kind but not in a numerical degree. Quantitative: Responses are finite numbers that can vary in magnitude.
Variance	The sum of the squared deviations of each individual value from the mean divided by the number of observations minus one for samples or just the number of observations for a population.

4.3 PROBLEM STATEMENT, IDENTIFICATION OF VARIABLES, AND HYPOTHESES

In Chapter 2, we explored the literature review process to gain knowledge as part of the process to initiate a research project. The literature review helps identify knowledge gaps, unresolved issues, and areas needing further investigation. Once the literature review is completed, the scientist is ready to define the specific goal of the project and is able to establish a problem statement, as well as a hypothesis or hypotheses to be evaluated via experiments.

- Problem Statement

  A "Problem Statement" is a concise description of an issue or lack of knowledge to be investigated. It defines the boundaries of the specific problem (e.g., existing condition), provides context to its significance (e.g., summary of knowledge available on the subject), and serves as justification for the study (e.g., what remains unknown and what will be the future state once the investigation is completed). A brief web search (e.g., WordNest.com or ResearchMethod.Net) will provide additional details regarding the development of problem statements. The problem statement, together with the hypothesis, is for developing the experiments as part of the investigation.

- Defining the Variables

  The next step, which also uses the outcome of the literature review, is defining the variables to be considered as part of the experimental work. The scientist should list all potential variables (e.g., independent and dependent), considering the information from previous research as well as his or her own ideas. **Table 4.1** summarizes the general definitions for the variables as well as some general examples. In some cases, a dependent variable may be considered as an independent one if the scientist decides to control such a variable as part of the investigation (e.g., time can be pre-defined as independent or can be measured as outcome). A key principle always in mind and defined by D. J. Wheeler (2000) is that "No data have meaning apart from their context."

The researcher must identify the variables, as previously mentioned, and assess how to control the independent variables and how to measure

TABLE 4.1 Definition of Variables

Type of Variables	Definition	Examples
Independent	An independent variable is the variable that a scientist manipulates or controls in an experiment to test its effect on dependent variables. The independent variable represents the cause in a cause-and-effect relationship and does not depend on other variables in the test. These variables are also called "factors" when working on the design of experiments.	Temperature, pressure, quantities, flow rates, electric filed intensity, number of pulses, retention times, treatment time.
Dependent	A dependent variable is the variable affected by independent variables. It represents the effect of an independent variable. These variables are also called "outcomes" or "observations."	Color, viability of cells, required time to complete a step.

the effect on the dependent variables. This will help define the number of experiments to be executed as part of the investigation, as well as the number and nature of samples that will be required. **Table 4.2** lists a few specific examples of independent and dependent variables associated with different topics.

Once the variables are defined, together with the problem statement, the scientist is ready to establish the hypothesis or hypotheses to be evaluated when executing the experiments.

- *Development of the Hypothesis*

  Hypotheses are statements or proposed explanations made as a starting point for further investigation. Once a hypothesis is postulated, it must be proven and verified. The following are examples of hypotheses developed in the science and engineering fields:

  Engineering:
  - The flow rate of cold water in a 1" i.d. PVC pipe is faster than the flow rate of cold water in a 1" i.d. galvanized iron pipe when comparing two pipes of equal length, constant temperature, and constant differential pressure between the two ends of the pipes.
  - The acceleration from 0 MPH to 60 MPH using an 8-cylinder, 500 hp engine is greater than the acceleration using a 4-cylinder, 150 hp engine, considering the same car body, wheels, and overall weight of the vehicles.

TABLE 4.2 Example of Independent and Dependent Variables

Topic	Independent Variables	Dependent Variables	Comments
Microbiology – Growth	• Type of microorganisms • Temperature • Concentration of nutrients • Presence of inhibitors • Time	• Number of colony forming units (CFU) • Morphology of cells	Health of the cells
Cell Therapies – Blood collection	• Temperature • T-cell collection protocol • Hold time post collection • Quantity of anticoagulants	• Number of CD4+ and CD8+ T-cells • Viability of cells • Residual RBC and platelets	Quality of T-cells intended for a CAR T product
Homogenization – Suspension of particles	• Pressure • Temperature • Flow rate	• Degree of separation or precipitation of particles	Quality of the resulting solution
Pasteurization – Inactivation of microorganisms	• Temperature • Residence time • Flow rate	• Viable microorganisms • Type of microorganisms • Shelf life of the product • Organoleptic characteristics (odor, color, flavor)	Effectiveness of the heat treatment
Non-thermal treatments – Pulse electric fields	• Temperature • Residence time • Flow rate • PEF intensity • PEF frequency • PEF number of pulses	• Viable microorganisms • Type of microorganisms • Shelf life of the product • Organoleptic characteristics (odor, color, flavor)	Effectiveness of the treatment
Freezing process – Preservation of cells	• Temperature • Volume of material • Freezing rate	• Time • Viable cells • Injured cells • Morphology of cells	Quality of the preserved materials

- The inactivation or killing of *Escherichia coli* suspended in whole milk at 25°C while using pulsed electric fields is more effective at 50 kV/cm at 100 Hz than at 25 kV/cm at 200 Hz at a constant flow rate of 100 mL/min.

Biology:

- The growth phase of *Staphylococcus aureus* is limited by the presence of garlic extract on a Petri dish.
- The absence of direct sunlight prevents the growth of grass under a bank of solar panels.
- T-cells will multiply faster when incubated at 36.0°C vs. incubation at 25°C.

These hypotheses can be expressed in mathematical terms for statistical purposes, considering the use of mean values (μ, mu) and standard deviations (σ, sigma). We will further discuss these later in the book.

4.4 BASIC STATISTICS

Any scientific research or investigation needs a hypothesis to set the stage for proving or disproving an issue. Once in place, the hypothesis can be translated to mathematical terms using mean and standard deviation values. Nevertheless, there are a few additional terms that scientists must understand. Let's start with the terms "Population" and "Sample." The definition of these terms is below and can be found elsewhere (Montgomery, 1984; Ott, 1993; Kiemele et al., 2000; Frost, 2025):

- **Population**: The whole group of items or elements of interest. It represents all the possible measures the scientist can take for a particular variable (e.g., size, color, weight, distance, etc.). The measures taken from a population represent the universe, and therefore any conclusion is about the whole population. When referring to measures taken from a population, the collected measures for a variable are known as parameters.
- **Sample:** A sample is a subset of the population that scientists select and include in their study. It is a limited number of measures from a population. These measures are randomly selected from the general population to fairly represent the population of the subjects or items to be evaluated. This is especially practical when studying very large populations where it is not possible to measure each subject or

element. Since a sample is a fraction of the population, samples must be representative so any conclusion reached from the samples can be assumed to apply to the overall population. When referring to measures taken from a sample, the collected measures for a variable are known as statistics.

Statistical tools are commonly used to assess scientific data compiled during investigations or studies. Statistics is the study of making sense of data (Ott, 1993). Let's start the discussion on statistics and statistical tools by defining measures of central tendency and measures of variability.

- Measure of Central Tendency

  Measures of central tendency are reported as (1) mean or arithmetic mean; (2) median; or (3) mode. **Table 4.3** summarizes the characteristics of each measure.

  The determination of the measure of central tendency is described in **Table 4.4**.

TABLE 4.3 Characteristics of Main Measure of Central Tendency

Measure of Central Tendency	**Characteristics**
Arithmetic mean or mean	- Average of the measures in a data set. - Only one value per data set. - Affected by extreme values. - Averages of subsets can be used to estimate the overall average. - Only applies to quantitative values.
Median	- It is the central point, 50% of measures are above and 50% are below this value. - Only one value per data set. - Not affected by extreme values. - Only applies to quantitative values. - Value is relatively stable even when data may be grouped in different subsets.
Mode	- It is the most frequent value from a data set. - More than one mode can be in a data set. - Not affected by extreme values. - Modes from subsets cannot be used to estimate the mode of a data set. - Mode may be changed by modifying data subsets. - Apply to both quantitative and qualitative data sets.

Source: (Ott, 1993).

TABLE 4.4 Estimate of the Measure of Central Tendency

Measure of Central Tendency	Estimate or Determination
Arithmetic mean or mean	The sum of all the measures divided by the number of observations. The mean for a population is expressed as mu (μ) while the value for a sample is expressed as ȳ. The calculation for μ follows the same formula as for ȳ. In the formula, y_i represents each individual value or measure, and n is the total number of observations. $\bar{y} = \frac{\sum_1^n y_i}{n}$
Median	The median is the middle value (50% are below and 50% are above) of the observations when placed in ascending order. In the case of an even number of observations, the median will be estimated by averaging the two values around the center of the data set.
Mode	The mode is the value that is repeated more often in a data set. In the case of multiple values with the same repeated frequency, all the values are considered "mode."

The following example shows the estimation or calculation of central tendencies.

In a class of 40 students, the height of 15 students was measured in a random manner. The height (meters) of the 15 students was: 1.5, 1.8, 1.4, 1.7, 1.8, 1.7, 1.4, 1.3, 1.4, 1.5, 1.4, 1.6, 1.7, 1.4, 1.5

Mean: Sum of the 15 data points = 23.1 and the mean = 1.5 m (23.1/15)

Median: Aligning the data in ascending order: 1.5 is the middle value.

1.3 1.4 1.4 1.4 1.4 1.4 1.5 **1.5** 1.5 1.6 1.7 1.7 1.7 1.8 1.8

Mode: Based on the frequency of observed values, 1.4 is the mode.

Value	Observed Frequency
1.3	1
1.4	**5**
1.5	3
1.6	1
1.7	3
1.8	2

- *Measures of Variability*

Together with measures of central tendency, the use of measures of variability or dispersion is important to characterize the population or the sample and to describe the uniformity of the collected data.

The measures of variability or dispersion most commonly used are range, percentiles, variance, and standard deviation. These measures are key for comparing the results from experiments (e.g., comparing different experimental conditions). **Table 4.5** summarizes the definition of each measure of variability.

TABLE 4.5 Definitions for Measures of Variability

Measure of Variability	Definition
Range	Range is the difference between the extreme values (e.g., largest and smallest) observed in a data set. This value allows a rapid assessment of the uniformity of the information in the data set.
Percentile	The nth percentile is defined as the value at which a percentage (p%) of the data is below the value and 100-p% is above the value. Commonly used percentiles are 25 (lower quartile), 50 (middle quartile), and 75 (upper quartile). Nevertheless, other percentages can be used to characterize a sample or population, considering the basic definition.
Variance	The variance of a set of measurements is the sum of the squared deviations of each individual value (y_i) from the mean ($\bar{y}$) divided by the number of observations (n) minus one (1) in the case of a sample and n in the case of a population: $s^2 = \frac{\sum_i (y_i - \bar{y})^2}{n-1}$ (for sample) $\sigma^2 = \frac{\sum_i (y_i - \bar{y})^2}{n}$ (for population)
Standard deviation	It is the positive square root of the variance. It is used "s" for denoting the sample standard deviation and σ for denoting the population standard deviation.

Source: (Ott, 1993)

The following example, previously used to estimate central tendency, shows the calculation or estimation of the measure of variability.

In a class of 40 students (population), the height of 15 students (sample) was measured in a random manner. The height (meters) of the 15 students was: 1.5, 1.8, 1.4, 1.7, 1.8, 1.7, 1.4, 1.3, 1.4, 1.5, 1.4, 1.6, 1.7, 1.4, 1.5

Range: min = 1.3 m max = 1.8 m *Range* = 0.5 m

Percentiles: Lower Quartile (25%): 1.4 Upper Quartile (75%): 1.7

Variance: $s^2 = 0.03$ m² (*Sum of squared differences* = 0.38); $n = 15$)

Standard deviation: 0.16 m

By combining the measure of central tendency and variability (e.g., mean and standard deviation), the scientists can make comparisons between two or more data sets of measurements. Also, the use of the standard deviation while considering the Empirical Rule (given a set of n points possessing a mound-shaped histogram or normal distribution) allows predicting the percentage of data points under the normal curve. This is summarized in **Table 4.6.**

TABLE 4.6 Empirical Rule Considering a Normal Distribution of Data

Interval	% of Data Points within the Interval
$\bar{y} \pm s$	68
$\bar{y} \pm 2s$	95
$\bar{y} \pm 3s$	99.9

Figure 4.1 presents the above information as a plot using an arbitrary scale, where $\bar{y}$ is 0.5 units and s is 0.1 units (general example for demonstration of the empirical rule only):

Ott (1993) indicates that the Empirical Rule is quite accurate in predicting the percentage within 2s in most data sets, while it may be less accurate in forecasting the percentage within one standard deviation of the mean. Also, considering that the value for range is estimated using the extreme values of a data set, the value for standard deviation can be approximated by dividing the range by 4 (Ott, 1993).

We have discussed the terms population and sample in the context of the number of observations that can be available for analysis (e.g., the whole set or a subset). Nevertheless, data can be classified in terms of the nature of the variables (e.g., Discrete and Continuous) and their characteristics (e.g., Qualitative and Quantitative). **Table 4.7** summarizes the types and definitions of these terms.

- *Random Sampling*

A key consideration when talking about variables in a scientific investigation is the need for random sampling while collecting data. This means that the value obtained for a variable is by chance and not biased (e.g., pre-disposed outcome). Specifically, the scientists must know how a sample is selected to assess the probability associated with various sample outcomes. The probability is then used to make inferences about the population from which the sample was taken. The selection of data in a non-random fashion will result in

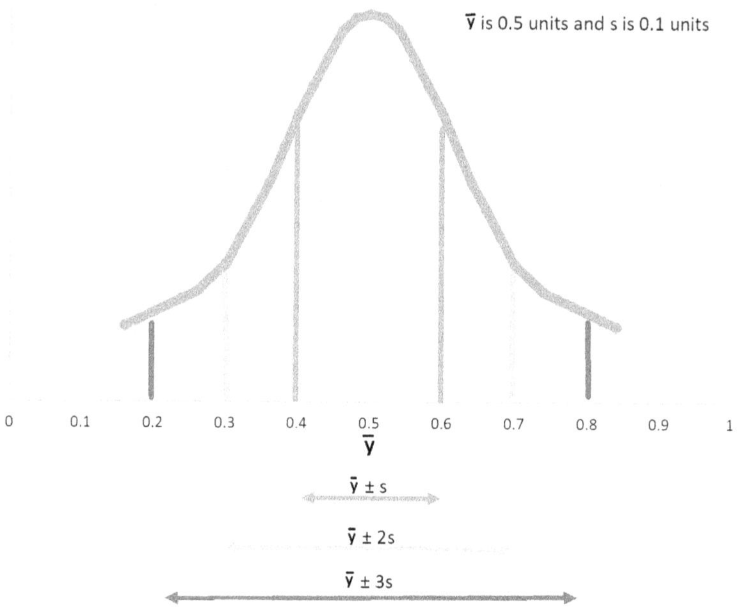

FIGURE 4.1 General example for representation of the empirical rule.

TABLE 4.7 Type and Characteristics of Variables

Type of Variable and Characteristics	Definition
Discrete	A variable with a finite number of values which be assumed when collecting data. Examples of discrete variables are the following: number of cars in a lot, number of accidents on a highway, number of votes for a candidate, number of metal pieces created by a machine.
Continuous	A variable with an infinite number of values that can be assumed in an interval. Examples of continuous variables are temperature, pressure, height, weight, distance, and speed.
Qualitative	Responses that can vary in kind but not in any numerical degree.
Quantitative	Responses are finite numbers, and those responses vary in numerical magnitude.

distorted or biased data. The bias can either overrepresent or underrepresent portions of the population. To ensure a sample is selected in a random fashion, the scientist must ensure that every different sample of size "n" from a population has the same probability of being selected. Ott (1993) highlights in his book that "data from a random sample will provide the foundation for making statistical inferences" and "random samples are not easy to obtain, but with care we can avoid many potential biases."

- *Central Limit Theorem*

Also, most variables have mound-shaped frequency distributions that can be approximated by using a normal curve or distribution that has been introduced previously when we discussed the Empirical Rule; in fact, the probabilities associated with a normal curve are the basis for the rule. Most of the methods to assess or compare experimental data will require those results to satisfy or follow a normal distribution. This leads to a discussion of one of the *Central Limit Theorems* that explains why most sampling distributions are normal.

Figure 4.2 presents the concept of the Central Limit Theorem described below:

> *If random samples of n measurements are repeatedly drawn from a population with a finite mean μ and a standard deviation σ, when n is large, the relative frequency histogram for the sample means will be approximately normal (bell-shaped) with mean μ and standard deviation $\frac{\sigma}{\sqrt{n}}$ (known as standard error of $\bar{y}$, this approximation becomes more precise as n increases).*

This is true even if the samples are taken from a population that is not normally distributed. In the case of **Figure 4.2 (a)**, the plot shows the individual values that may not fall under a normal distribution. Meanwhile, **Figure 4.2 (b)** shows how the distribution of the means from multiple random samples taken from that population follows a normal distribution.

Additional information on this subject is found elsewhere (e.g., Montgomery, 1984; Stell & Torrie, 1980; Ott, 1993).

4.5 DESIGN OF EXPERIMENTS

Experimental work can be conducted considering different designs of experiments such as *Completely Randomized Design, Randomized Block Design, Latin Square Design, and Factorial Designs.* We strongly suggest

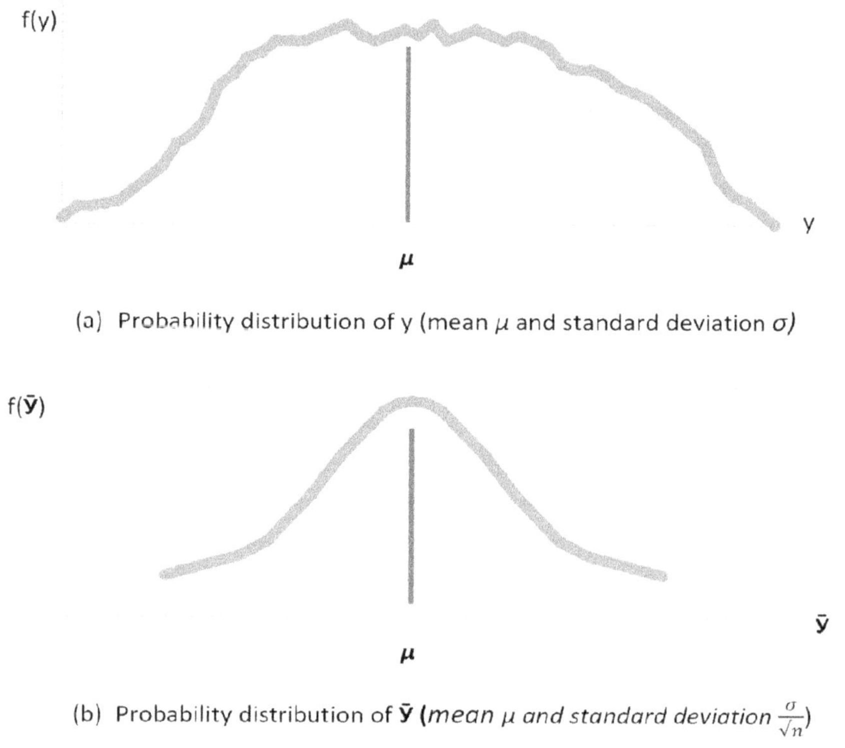

FIGURE 4.2 Probability distribution of y and the sampling distribution of $\bar{y}$.

the readers further review these concepts in books specifically focused on statistics and these designs, as well as additional designs (Stell and Torrie, 1980; Montgomery, 1984; Ott, 1993).

- *Completely Randomized Design*

  It is the experimental setting where a random sample of "n" observations is collected from each of "t" different populations. Data for a *Completely Randomized Design* can be summarized as shown in **Table 4.8**.

TABLE 4.8 Completely Randomized Design – Data and Statistics

Population	Data					Total	Mean
1	y_{11}	y_{12}	y_{31}	–	y_{1n}	y_{1-}	$\bar{y}_{1-}$
2	y_{21}	y_{22}	y_{32}	–	y_{2n}	y_{2-}	$\bar{y}_{2-}$
3	y_{31}	y_{32}	y_{33}	–	y_{3n}	y_{3-}	$\bar{y}_{3-}$
–	–	–	–	–	–	–	–
t	y_{t1}	y_{t2}	y_{t3}	–	y_{tn}	y_{t-}	$\bar{y}_{t-}$

where y_{ij} represents individual observation or sample ($j = 1$ to n) in each population ($i = 1$ to t); $y_{i\cdot}$ represents the sum of the sample measurements obtained for population "i"; and $\bar{y}_{i\cdot}$ represents the mean value for population "i."

This information will be used later when we introduce the use of Analysis of Variance (ANOVA) for the statistical analysis of the data when comparing two or more populations.

- *Randomized Block Design*

This is an extension to the completely randomized design where a potential source of variation is "blocked" (e.g., analysts in a testing lab, multiple testing pieces of equipment, testing times, or dates). The design is used to compare "t" different treatment means when an extraneous source of variability (e.g., blocks) is present. The idea is to run each treatment within each block to remove the variability between blocks. The following diagram shows the concept of blocks during an experiment. On the left, we have all testing for each treatment (A, B, and C) assigned in a random way to a particular piece of equipment. This arrangement will not allow us to establish if any difference between A, B, and C is the result of the treatments or introduced by the testing equipment. Meanwhile, by creating three blocks (1, 2, 3) and randomly testing a sample of each treatment under each block (right side of **Table 4.9**), the issue is resolved.

TABLE 4.9 Randomized Block – Experiment Design

CRD			**Block**		
Test Equipment			**Test Equipment**		
1	2	3	1	2	3
A	B	C	A	B	A
A	B	C	B	A	C
A	B	C	C	C	B

Data for a *Randomized Block Design* can be summarized as shown in **Table 4.10**.

TABLE 4.10 Randomized Block – Data and Statistics

	Blocks						
Treatment	**1**	**2**	**3**		**b**	**Total**	**Mean**
1	y_{11}	y_{12}	y_{13}	–	y_{1n}	$y_{1\cdot}$	$\bar{y}_{1\cdot}$
2	y_{21}	y_{22}	y_{23}	–	y_{2n}	$y_{2\cdot}$	$\bar{y}_{2\cdot}$
3	y_{31}	y_{32}	y_{33}	–	y_{3n}	$y_{3\cdot}$	$\bar{y}_{3\cdot}$
–	–	–	–	–	–	–	–
t	y_{t1}	y_{t2}	y_{t3}	–	y_{tb}	$y_{t\cdot}$	$\bar{y}_{t\cdot}$
Totals	$y_{\cdot 1}$	$y_{\cdot 2}$	$y_{\cdot 3}$	–	$y_{\cdot b}$	$y_{\cdot\cdot}$	
Means	$\bar{y}_{\cdot 1}$	$\bar{y}_{\cdot 2}$	$\bar{y}_{\cdot 3}$	–	$\bar{y}_{\cdot b}$		$\bar{y}_{\cdot\cdot}$

where y_{ij} represents individual observation for treatment "*i*" in block "j"; "*t*" is the number of treatments; "*b*" is the number of blocks; $y_{i\cdot}$ represents the sum for treatment "i"; $y_{\cdot j}$ is the sum for all observations in block j; $y_{\cdot\cdot}$ represents the sum of all observations; $\bar{y}_{i\cdot}$ represents the mean value for treatment "i"; $\bar{y}_{\cdot j}$ represents the mean value for block "j"; and $\bar{y}_{\cdot\cdot}$ represents the overall sample mean.

This information will be used later when we introduce the use of Analysis of Variance (ANOVA) for the statistical analysis of the data when comparing two or more treatments.

- *Latin Square Design*

A Latin Square Design contains *r* rows and *c* columns, where *r* and *c* are equal in magnitude. Treatments are randomly assigned to experimental units within the rows and columns so that each treatment is represented in every row and in every column. Using the example from the Randomized Blocks, if a second extraneous factor is introduced that may affect the treatments (e.g., experiment or testing on different days), as shown in the diagram on the right, there is a need to separate the impact of that second factor from the actual treatments. This is attained by ensuring each treatment appears in each row (Day) and each column (Test Equipment) as shown in **Table 4.11**.

TABLE 4.11 Latin Square – Experiment Design

Block			Latin Square			
Test Equipment				Test Equipment		
1	2	3	**Day**	1	2	3
A	B	A	1	A	C	B
B	A	C	2	B	A	C
C	C	B	3	C	B	A

Data for the *Latin Square Design* can be summarized as shown in **Table 4.12** for a 3 × 3 design:

TABLE 4.12 Latin Square – Data and Statistics

	Column				
Row	**1**	**2**	**3**	**Total**	**Mean**
1	y_{111}	y_{312}	y_{213}	y_{-1-}	$\bar{y}_{-1-}$
2	y_{221}	y_{122}	y_{323}	y_{-2-}	$\bar{y}_{-2-}$
3	y_{331}	y_{232}	y_{133}	y_{-3-}	$\bar{y}_{-3-}$
Totals	y_{-1}	y_{-2}	y_{-3}	y_{---}	
Means	$\bar{y}_{-1}$	$\bar{y}_{-2}$	$\bar{y}_{-3}$		$\bar{y}_{-}$

where y_{ijk} represents the individual outcome for treatment "i" in row "j" and column "k"; t is the number of treatments as well as the number of columns and rows; y_{i--} represents the sum of the outcomes for treatment "i" across all columns and rows; $\bar{y}_{i--}$ represents the mean value for treatment "i" across all columns and rows; y_{-j-} is the sum of outcomes in row "j"; $\bar{y}_{-j-}$ represents the mean value of outcomes in row "j"; y_{--k} is the sum of outcomes in column "k"; $\bar{y}_{--k}$ represents the mean value of outcomes in column "k"; y_{---} represents the sum of all outcomes; and $\bar{y}_{---}$ represents the overall mean.

This information will be used later when we introduce the use of Analysis of Variance (ANOVA) for the statistical analysis of the data when using a Latin Squares Design.

- *Factorial Design*

A Factorial Design is an experiment in which a response is observed at all combinations of the independent variables. These are useful when examining the effects of two or more factors (independent variables) on a response or outcome "*y*" (dependent variable). The choice of the number of levels of each variable and their settings is important. Also, random assignment of the treatments to experimental units is critical. This type of design allows evaluating the effect of multiple variables while avoiding one-at-a-time experiments, which could result in an incorrect inference when a scientist assumes that the effect of one factor is the same for multiple levels of other factors. **Figure 4.3** presents the potential outcome "*y*" of a Factorial Design considering two factors (A and B). On the left side, Factors A and B are said to interact while on the right Factors A and B do not interact. An interaction represents the influence of one factor on another.

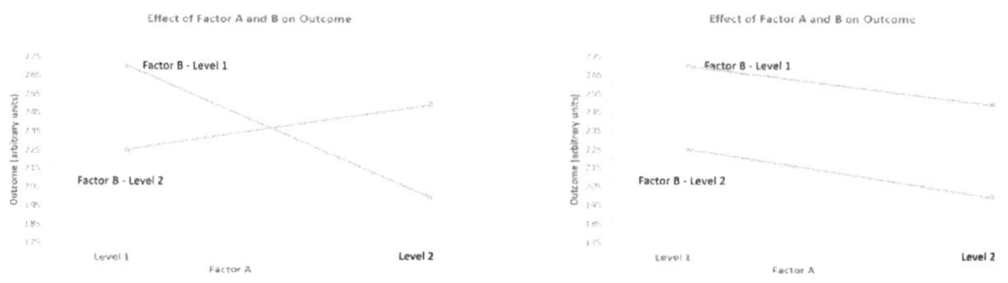

FIGURE 4.3 Outcomes of "y" of a factorial design considering two factors (A and B).

The number of factors is defined as part of developing the experiments. The complexity of the statistical analysis will increase as the number of factors is increased. **Table 4.13** summarizes the data arrangement for a Two-Factor Factorial Design with "n" number of samples in each experiment. Also, the number of levels for each factor can be different.

TABLE 4.13 Factorial – Experimental Design

			Factor B		
		1	2		b
	1	y_{111} y_{112} y_{11n}	y_{121} y_{122} y_{12n}	–	y_{1b1} y_{1b2} y_{1bn}
Factor A	**2**	y_{211} y_{212} y_{21n}	y_{221} y_{222} y_{22n}	–	y_{2b1} y_{2b2} y_{2bn}
	–	–	–	–	–
	a	ya_{11} ya_{12} ya_{1n}	y_{a21} y_{a22} y_{a2n}	–	y_{ab1} y_{ab2} y_{abn}

This information will be used later when we introduce the use of Analysis of Variance (ANOVA) for the statistical analysis of the data when using a Factorial Design.

The advantages and disadvantages of each of the different designs are summarized in **Table 4.14**. These must be considered as part of the decision process while preparing experiments.

4.6 SAMPLES AND SAMPLING PLANS

The collection of data during an experiment must follow a well-defined process to ensure the quality and integrity of the information. Also, the data shall be generated according to the experimental design intended for the experiment, so the analysis and conclusions are valid and free of any biases.

TABLE 4.14 Advantages and Disadvantages for Designs of Experiments

Type of Design	Advantages	Disadvantages
Completely randomized	• Easy to construct. • Easy analysis, even with different sample sizes. • Can be used for any number of treatments (e.g., populations)	• Best suited for cases with few treatments. • Homogeneity of the test items is critical to avoid extraneous sources of variability.
Randomized block	• Practical when comparing multiple treatments with a single extraneous source of variability. • Simple statistical analysis. • Flexible for any number of treatments in any number of blocks.	• Best suited for a small number of treatments. • Design focuses on single extraneous source of variability. • Effect of treatment must be about the same from block to block.
Latin square	• Appropriate for comparing *t* treatments means in the presence of two sources of extraneous variation. • Relatively easy analysis.	• Most practical for a number of treatments between 5 and 10. • Additional sources of variability (>2) increase the error. • Effect of treatment on outcome must be about the same across rows and columns.
Factorial	• Reduce the potential error of inferring an outcome for a particular combination of factors. • Allows determination of interactions between factors. • Number of levels per factor is independent. • Number of replicates can vary.	• A more complex analysis is required. • Key attention required when randomizing the testing of factors.

Source: (Montgomery, 1984; Ott, 1993)

The following elements should be considered while defining, collecting, and testing samples required for any experiment:

- Source and quantity of samples should be clearly defined in the study protocol. Spare or back-up samples may be considered as part of the sampling activities.
- Collection of the samples should use techniques, containers, and equipment that will not alter the composition or otherwise damage the sample.

- Each sample shall be properly identified, including content, sampling date/time, storage condition, and collector.
- Samples shall be protected from environmental conditions that could damage or modify the content (e.g., sunlight, UV light, extreme temperature conditions, chemical contaminants).
- Samples requiring refrigeration or freezing should be transferred to the storage area as soon as possible and "time out of refrigeration" data shall be compiled as part of the lifecycle of the sample.
- Pre-testing handling of any sample shall be documented as part of the lifecycle of the sample.
- Testing procedures for collected samples shall describe all the requirements for valid results.

The number of samples collected during an experiment will affect the quality of the inferences intended once the experiment is fully executed and data are completely analyzed. In general, the larger the number of samples, the stronger any inference will be. Nevertheless, there is small a chance of having an infinite number of samples due to elements such as: (1) quantity of materials available for testing; (2) amount of time required for testing samples; (3) type of analytical testing required for samples; (4) number of analytical tests per sample (e.g., different tests); (5) nature of the experiment.

The minimum number of samples required for any experiment can be estimated considering the type of design of experiment and associated statistical analysis. In some cases, only a single observation (or replicate) per experimental condition is possible, and the associated inference and statistical analysis must clearly state such a constraint. Replication is always recommended to properly assess the impact of factors or independent variables on dependent variables.

The scientists should prepare a detailed sampling plan as part of the experiment. The plan should provide clarity around the number of samples to be collected, the timing for those samples, their handling, and their storage.

4.7 LABORATORY WORK

The experimental work is not limited to the design and execution of the experiments. The scientists must consider the facility in which the work will be conducted, including the environmental conditions, safety considerations, analytical methods, and analytical instruments.

4.7.1 Facility

The laboratory where experimental work will be conducted must satisfy the local, state, and federal regulations. Appropriate work areas (e.g., benches, waste disposal systems, raw material and chemical storage) and pieces of equipment shall be available to ensure satisfactory execution of the experiments.

4.7.2 Environmental Conditions

The facility shall be clean and proper environmental controls in place (e.g., chemical hoods, biological safety cabinets, working air conditioning systems, and clean benches and floors).

4.7.3 Safety Considerations

The use of chemicals and biological compounds and agents requires proper knowledge of the potential health and environmental risks associated with them. The scientists shall have full access to *Safety Data Sheets* (SDS) where information related to the risks and how to react in case of an incident or accident involving chemicals or biological agents.

The design and operation of research laboratories in the USA are regulated by different agencies (e.g., Food and Drug Administration – FDA; the Occupational Safety and Health Administration – OSHA; the Environmental Protection Agency – EPA; and the Department of Transportation – DOT). The main objective of the regulations is to ensure that people, surrounding neighborhoods, and the natural resources in the area are not impacted by improper operation of a facility.

4.7.4 Analytical Methods and Instruments

Experimental work may require analytical testing to characterize a sample. The complexity of analytical testing is associated with the nature of the tests: for example, simple pH measurement versus a complex composition analysis using High Pressure Liquid Chromatography (HPLC) or Gas Chromatography (GC). In any case, the scientists shall consider the following:

- Detailed procedures for handling and processing samples for each analytical method required for the experiment.
- The scientist conducting analytical testing should be fully trained in those methods as well as the operation of the equipment.

- Each piece of equipment shall be calibrated and set for the intended type of samples (e.g., pressure ranges within limits of a column, HPLC mobile phase compatible with the sample).
- Pre- and post-use verification of the instrumentation to ensure proper operation before and after its use.
- Proper preparation and use of control samples.
- Equipment should be properly installed and qualified.

4.8 FINAL COMMENTS

The relevance and considerations of the design of experiments have been highlighted in this chapter. Proper selection of the independent variables and the dependent variables is linked to the successful execution of the experiment, as well as the data analysis following data collection.

4.9 REFERENCES

Frost .J 2025. Population vs Sample: Uses and Examples - Statistics By Jim. https://statisticsbyjim.com/basics/population-vs-sample/.

Kiemele, M.J., Schmidt, S.R., and Berdine, R.J. 2000. *Basic Statistics: Tools for Continuous Improvement.* 4th Edition. Air Academy Press, LLC, Colorado Spring, CO.

Montgomery, D.C. 1984. *Design and Analysis of Experiments.* 2nd Edition. Joh Wiley & Sons, Inc., New York, NY.

Ott, L. 1993. *An Introduction to Statistical Methods and Data Analysis.* 4th Edition. Duxbury Press, Belmont, CA.

ReseachMethod.Net. 2022. Problem Statement — Writing Guide, Examples and Types. https://researchmethod.net/problem-statement/#google_vignette.

Stell, R.G.D., and Torrie, J.H. 1980. *Principle and Procedures of Statistics. A Biometrical Approach.* 2nd Edition. McGraw Hill, Inc., New York, NY.

Wheeler, D.J. 2000. *Understanding Variation: The Key to Managing Chaos.* 2nd Edition. SPC Press, Knoxville, TN.

WordNest. 2026. Problem Statement: Definition, Components, Examples & How-To Guide. https://wordnest.org/problem-statement/.

CHAPTER 5

Data Collection and Analysis

5.1 INTRODUCTION

Data collection and analysis are the core elements of scientific research because data are the facts leading to a conclusion. In this chapter, we provide the readers with guidance on how to collect and analyze the results of experiments. Among the topics covered are: laboratory notebooks, logbooks, preparation of data tables, graphics or plots, and statistical tests.

5.2 GLOSSARY OF TERMS

The following terms, commonly used by scientists, are used in this chapter.

Terms	Definition
Analysis of variance (ANOVA)	A family of statistical methods used to compare the means of two or more groups by analyzing variance. Specifically, ANOVA compares the amount of variation between the group means to the amount of variation within each group. If the between-group variation is substantially larger than the within-group variation, it suggests that the group means are likely different.
Degrees of freedom	It is the number of independent values in a statistical calculation that can vary without violating any constraints imposed on the data.

DOI: 10.1201/9781042010516-5

Terms	Definition
Error Type I and II	Define the probability of selecting the incorrect answer when testing a hypothesis.
Logbooks	Books with bound pieces of paper with numbered pages.
Notebooks	Books with bound pieces of paper with numbered pages.
Plots	A graph showing the relation between two or more variables. Examples include *Pie Charts, Bar Charts, Histograms, Stem-&-Leaf.*
Rejection region	Statistical consideration to reject a hypothesis considering the Test statistic and tabulated standard values (e.g., *Z* values, *t* values, *F* values).
Tables	A set of facts or figures systematically displayed, especially in columns.
Test	Statistical method to assess a hypothesis.

5.3 DATA COLLECTION AND STATISTICAL ANALYSES

We have stated in previous chapters the basic requirements of collecting as much data as possible when conducting scientific research. All information shall be properly compiled and stored either in paper or digital notebooks. The information is later used when publishing the results in scientific journals or when presenting papers in scientific forums such as symposiums or conferences.

1. **Laboratory Notebooks and Logbooks**

   Use of paper or digital laboratory notebooks (sometimes referred to as logbooks) is essential for the collection of experimental raw data (e.g., literature review highlights, design of experiments, experimental conditions, observations, and test results). The integrity of the data is critical as supporting evidence of the research activity.

   In general, paper notebooks or logbooks are books with bound (stitched or glued) pieces of paper (sheets) with numbered pages. Each page may be blank, dotted, lined, or grid. The entries in a notebook or logbook should be made in permanent ink. The ink used for laboratory experiments is dark (e.g., black or blue), but other colors may be used to highlight particular elements or factors associated with the data. Today's technology allows making copies of original documents containing multiple colors. However, the best practice is the use of blue or black ink.

   In terms of general best practices, **Table 5.1** provides a summary of typical content of the notebooks.

TABLE 5.1 Typical Content of a Laboratory Notebook or Logbook

Page Number	Content	Comments
1	Owner: Phone number: Notebook number: Start date: Lab. location/office: Experimental work: End date:	This information highlights the ownership of the notebook or logbook. In case it is misplaced, any person finding the notebook shall be able to return it to the owner.
2	Table of contents	Chronological list of content captured in the notebook.
3 to (n–1)	Design of experiment: Scientists: Introduction: Variables: Methods: Equipment: Materials: Observations: Calculations:	Comprehensive information associated with the experimental work under development, including the bibliographic listings of the literature review, details of the experiments (e.g., equipment, methods, variables, test conditions), as well as observations and test results.
n to (p–1)	Design of experiment: Scientists: Introduction: Variables: Methods: Equipment: Materials: Observations: Calculations:	See the above comment.
p	Closing remarks for the notebook and a reference to the next notebook. Any blank page shall be crossed with a single line, dated and initialized.	This information serves as a link between other notebooks or logbooks.

The following items are part of good laboratory practices that the scientists shall use:

- Use of permanent ink (e.g., black, blue) when completing all required information in the notebook.
- Corrections of mistakes should be made by using a single strikethrough line over the affected entry, adding the correct information, adding an explanation for the correction, and applying the date and initials nearby the mistake to make clear who made the correction

and why the correction was needed. Do not erase or use "liquid paper" or any other correction fluid that makes the original data entry unreadable.

- The notebook or logbook shall be dedicated solely to the experimental work the scientist is conducting. The author should refrain from adding non-scientific information or extraneous comments to the notebook.

2. **Data Collection for Experimental Work**
The scientist should ensure the data collected during experimental work is properly tabulated and, when appropriate, presented in plots for easy analysis and interpretation. In this section, we provide general guidance on how to prepare tables and plots as part of the data collection and preliminary data analysis.

   a. Tables
   The use of tables is the primary option when collecting and presenting raw data. The following elements should be considered when preparing tables:
   - The title of the table must be self-explanatory. This will allow the reader to understand the goal of the table and its content.
   - All data will be captured in the table. In those cases of missing data, a note should be made explaining the reason for the missing information.
   - Avoid unnecessary information, complexity, redundancy, or inconsistency while collecting or presenting data. In the case of a constant variable or condition, it should not be listed multiple times in a table. Instead of repeating the same value, make a note highlighting the condition or list it in the title.
   - Avoid tables with too many columns or rows that make the table difficult to read. The format of a table should be adjusted to the needs of the experiment and the data being collected.
   - Each column and row in a table shall be properly identified, considering the nature of the data presented. Also, the identification should include the measuring units of the collected data.

Table 5.2 presents an example of a table summarizing the data compiled during a survey in a particular town of 125,000 inhabitants during the

TABLE 5.2 Per-Capita Annual Income at Town X in January 2023 (125,000 participants)

Per Capita Income (US $)	Number of Responses (Total)	Fraction of the Population (%)
< $20,000	15,000	12
$20,000 to $29,999	45,000	36
$30,000 to $39,999	40,000	32
$40,000 to $49,999	20,000	16
> $50,000	5,000	4

month of January 2023. The goal of the survey was to identify the distribution of per capita annual income in the town, considering all the people in a rural town "X." Five categories were used to define the incomes.

Table 5.3 presents an example of a table summarizing the data from a production line making plastic balls. The target weight of the ball is 4.0 pounds (lbs.).

TABLE 5.3 Weight (lbs.) of 36 Plastic Balls Manufactured during a Three-Hour Production Period at Factory W Using Two Molding Machines Running at Similar Process Conditions

Machine (ID)	Ball Weights (lbs)	General Observations
1	4.0, 3.0, 3.5, 4.5, 3.8 4.0, 4.2, 4.1, 4.0, 4.0 3.9, 3.9, 3.8, 4.0, 4.1 3.5, 4.3, 4.6	- Same lot of plastic (polypropylene) raw material used on both machines. - Production speed: 6 balls/hours - Three operators managed the two machines. - Temperature of molding machine: 200°C
2	4.5, 4.0, 3.9, 3.8, 3.3 3.2, 3.3, 3.7, 3.9, 3.9 4.1, 4.2, 4.2, 4.0, 3.8 4.0, 3.5, 4.3	

These tables can be part of the raw data compiled in the notebooks as well as later recreated in publications and presentations.

b. Plots

Plots are used to summarize compiled data. In this section, we will present as examples the use of pie charts, bar charts, histograms, stem-and-leaf plots, and time series.

(1) *Pie Charts*

This type of graphic representation (plot) is useful when a limited number of categories is considered for a variable. Typically, five or six categories are considered to ensure an easy and rapid interpretation of the results. The goal is to represent the potential alternatives of a variable. **Figure 5.1** presents the data discussed in Table 5.1, considering both the number of responses and the % population and presented in the form of a pie chart.

(2) *Bar Charts*

This type of graphic representation is useful for direct comparisons between data sets. Typically, the plot considers two axes: (1) the "x" axis listing a dependent variable with multiple potential outcomes and (2) the "y" axis presenting the quantification of each potential outcome. **Figure 5.2** presents data previously discussed in Table 5.1 and Figure 5.1. **Figure 5.3** is another example that summarizes the estimates of consumption per capita of frozen foods and cereals across different countries (note: data obtained from the internet with no standardized report for these data; this is intended as an example for the use of bar charts).

FIGURE 5.1 Representation of survey data (Table 5.1) using pie charts.

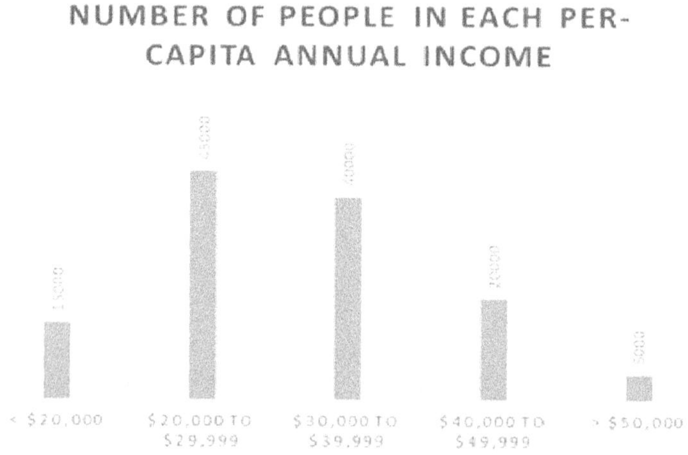

FIGURE 5.2 Representation of survey data (Table 5.1) using a bar chart.

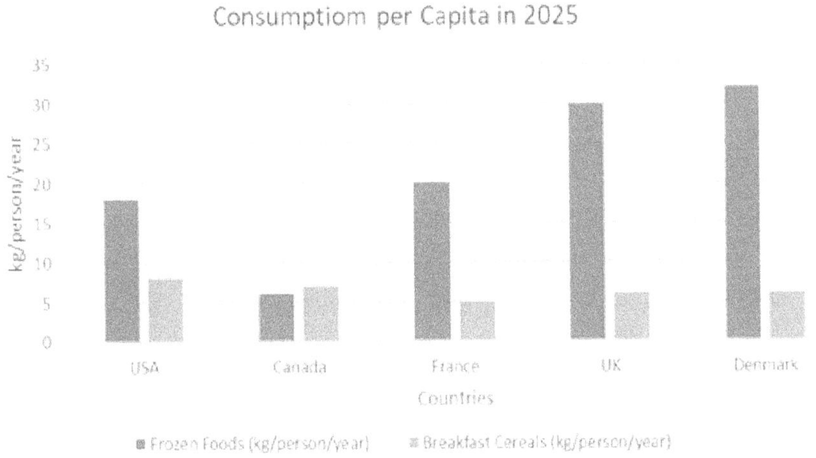

FIGURE 5.3 Estimated consumption per capita of frozen foods and cereals in 2025 across the USA, Canada, France, UK, and Denmark.

(3) *Histograms*

This type of graphical representation is typically used for quantitative, discrete, or continuous variables. The preparation of a histogram requires several steps after collecting the data:

- Determine the range of the measurements.
- Divide the range by the number of categories to be utilized, creating a series of intervals.
- Round the extreme values of each interval. Make sure not to have individual values on those extreme values.
- Quantify the frequency for each interval. Plot the frequency data per interval in the histogram.

The data presented in Table 5.3 is recreated in the form of a histogram in **Figure 5.4.** In this example, we have three plots: (1) All the data combined for both machines and (2) data for each machine.

The histogram can be used to report the cumulative frequency in addition to the individual frequencies as shown in **Figure 5.5** using the data from Table 5.3 and Figure 5.4.

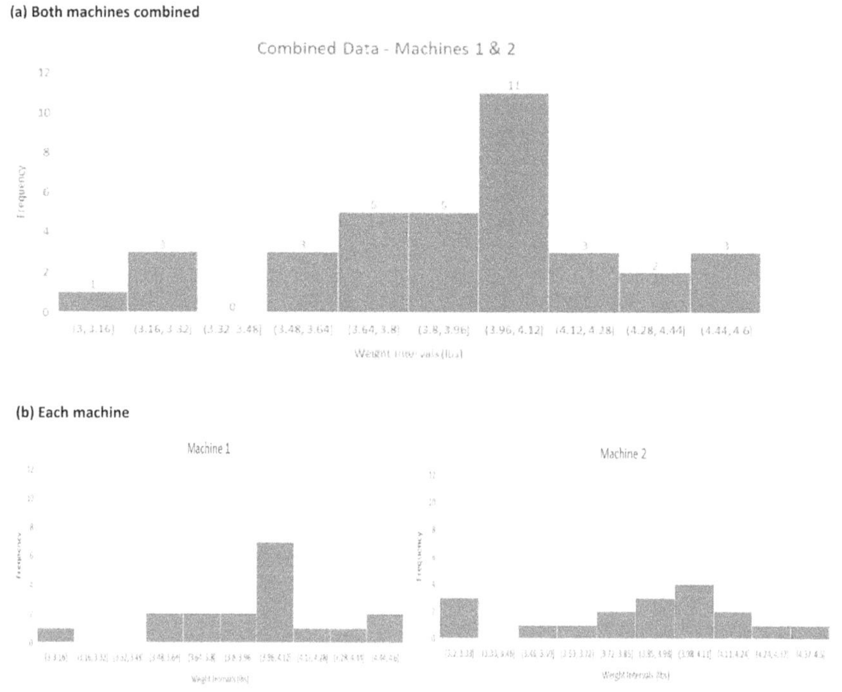

FIGURE 5.4 Weight (lbs.) of 36 plastic balls manufactured during a three-hour production period at Factory W using two molding machines running at similar process conditions.

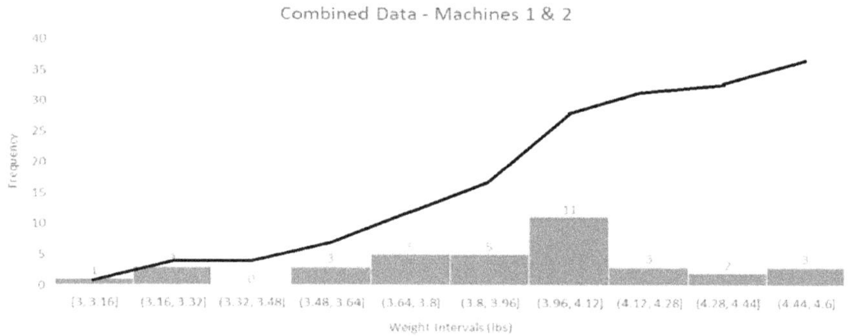

FIGURE 5.5 Cumulative frequency of ball weights (lbs).

(4) *Stem-and-Leaf*

This graphical representation considers ordering the "leaf" (trailing digits of a data point) along a row or "stem" (leading digits to the left) in ascending order. This allows counting the number of points in the stem value. It is an option for preparing a histogram-like picture of a frequency distribution. **Table 5.4** summarizes an example of the stem-and-leaf diagram for the following data set (survey about the number of books in each of 30 houses).

TABLE 5.4 Stem and Leaf Diagram for the Number of Books in 30 Houses

Stem	Leaf
1	10 20 23 35 41 42 45 45 47 65 65 68 78 87 89
2	00 10 15 22 30 36
3	10 15 54 58 69
4	12
5	21 26
6	
7	89

House	No. Books	House	No. Books	House	No. Books
1	200	11	412	21	135
2	215	12	526	22	142
3	315	13	165	23	120
4	369	14	141	24	147
5	145	15	189	25	168
6	123	16	210	26	110
7	178	17	236	27	187
8	222	18	358	28	165
9	789	19	354	29	521
10	145	20	310	30	230

The Stem ranges from 1 to 7 (data range from 110 to 789) because the digits on the left are 01 to 07.

(5) *Time Series*

Plots for time series allow scientists to assess general tendencies and cyclic patterns as a function of time (e.g., minutes, hours, days, weeks, months, years). The data is plotted in chronological order. Examples of these plots include company revenue, surveys, ticket sales, attendance at events, and sample test results (e.g., lot assays, microbial counts, pH, osmolality, weight, and chemical concentration).

Figure 5.6 presents the estimate of US Retail Sales for the year between January and November 2025 (note: data obtained from the internet with no standardized report for these data; this is intended as an example for the use of time series charts).

FIGURE 5.6 Estimated US retail sales in 2025.

3. **Data Analysis: Statistical Evaluations**
   This section provides the reader with a high-level summary of the concept of probabilities and statistical tests for data collected during experiments. Keep in mind what was previously discussed in Chapter 4 involving the design of experiments, estimates of mean and standard deviations (samples and populations), and the type of random variables (e.g., discrete and continuous).

   a. **Probabilities – Discrete Random Variables**
      The probability distribution of a discrete random variable y is denoted by $P(y)$. This distribution can be represented using a table, a mathematical expression, or a graph. As an example, consider tossing two coins. The possible outcomes for the number of heads are 0 (no heads), 1 (one head), or 2 (two heads). The probability distribution for this experiment is summarized in **Figure 5.7**.

There are three properties associated with probabilities involving simple discrete random variables:

- Probability associated with each value of y will be between 0 and 1.
- The sum of the probabilities for all values of y is equal to 1.
- Probabilities for discrete random variables are additive.

The example previously discussed is a classical binomial experiment. In this type of experiment, the scientist will repeat the same

y (Number of Heads)	P(y) – Probability of Heads
0	0.25
1	0.50
2	0.25
Totals	1.00

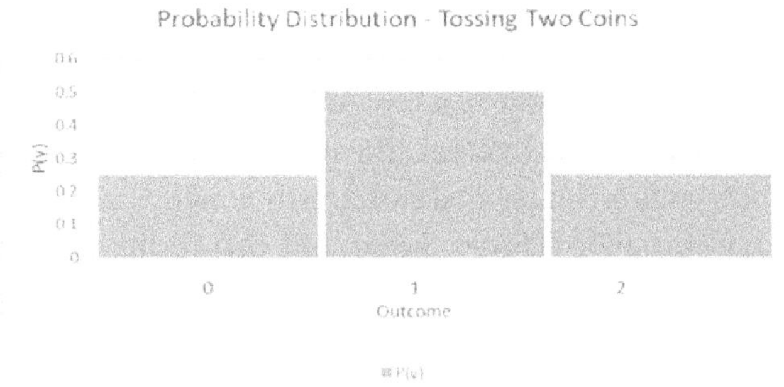

FIGURE 5.7 Probability distribution for the no. of heads – tossing two coins.

process n times (identical trials) and the result will be one of two outcomes (e.g., heads or tails, success or failure) each time. Each trial or experiment is independent (e.g., results from one trial do not influence the result of any other trial). The following formula can be used to predict the probability of "*y*" successes in "*n*" trials of a binomial experiment (Ott, 1993):

$$P(y)=\frac{n!}{y!(n-y)!}\pi^{y}(1-\pi)^{n-y}$$

where:

y is the number of successes in n trials.
n is the number of trials.
π is the probability of success on a single trial.
$1-\pi$ is the probability of failure on a single trial.
$n!=n(n-1)(n-2)\ldots.(3)(2)(1)$ (also known as n factorial).

The mean (μ) and standard deviation (σ) of the binomial probability distribution are estimated per the following formulas:

$$\mu = n\pi \qquad \sigma = \sqrt{n\pi(1-\pi)}$$

For more details on this topic, we recommend the review of the references listed at the end of the chapter.

b. **Probabilities – Continuous Random Variables**

For continuous random variables, the possible values span an entire interval or range. That is, the variable can take on an infinite number of values within a continuous interval. As a result, it is not meaningful to assign a finite probability to any single value of y. Instead, a relative frequency histogram is used to represent the probability that y falls within a given interval. The number of intervals can be large, and their widths can be made arbitrarily small.

The resulting plot **(Figure 5.8(a) and (b))** forms a smooth curve that represents the population's relative frequency distribution based on repeated observations of a continuous variable. The total area under the curve equals 1, representing the sum of all probabilities, as shown in **Figure 5.8(a)**. The probability that y lies within a specific interval between values i and j is given by $f(i < y < j)$, as illustrated in **Figure 5.8(b)**.

The probability distribution for the normal curve can be represented by the following equation:

$$f(y) = \left(\frac{1}{\sigma\sqrt{2\pi}}\right)e^{-0.5\left(\frac{y-\mu}{\sigma}\right)^2}$$

where:

π is 3.14159.

σ is the standard deviation.

μ is the mean.

y is the value of the random variable.

$f(y)$ is the probability of y.

As previously discussed in Chapter 4, 99% of the values under the curve occur within three standard deviations of the mean $(\mu \pm 3\sigma)$, 95% of the values are within two standard deviations $(\mu \pm 2\sigma)$, and 68% of the values are within one standard deviation. It is not practical to build individual curves for each potential combination of μ and σ, therefore the concept of **z** standard deviations with respect to

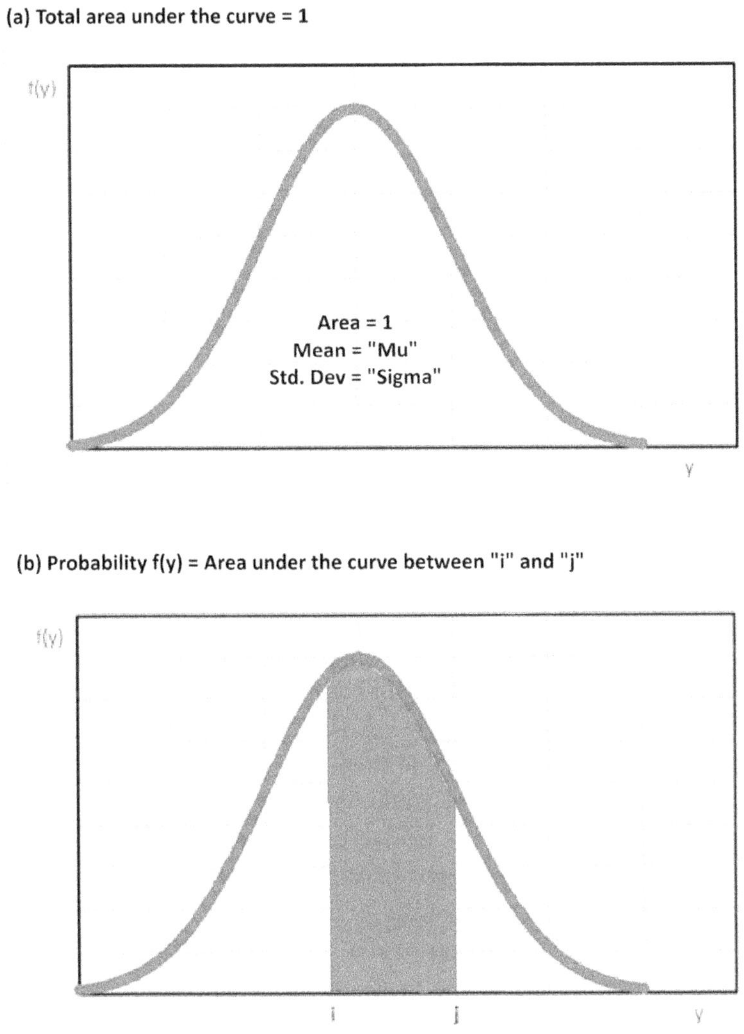

FIGURE 5.8 Probability distribution for a continuous random variable.

the mean is introduced for the representation of any potential combination of μ and σ. The value **z** is estimated using the following formula:

$$z = \frac{y - \mu}{\sigma}$$

This formula is used to estimate the probability, from the tabulated data for the cumulative standard distribution, of a value "y" in an interval

$(y - \mu)$. An example of the table is found elsewhere (Steel and Torrie, 1980; Ott, 1993; Montgomery, 1984; Kiemele et al., 2000). The normal curve is commonly utilized while evaluating experimental test results, specifically when comparing treatments and experimental conditions.

c. **Hypothesis Testing and Errors (Types I and II)**

Every experiment is guided by hypotheses that are evaluated once the experimental data are collected. Typically, a research or alternative hypothesis (H_a or H_1) is tested against a null hypothesis (H_0). After defining the hypotheses, the next step is to select an appropriate test statistic and determine the rejection region (that is, the criteria for rejecting H_0 in favor of H_a). Finally, a conclusion is drawn based on the statistical analysis.

The scientists must be aware of the two kinds of errors that may be committed when testing the hypotheses:

- **Type I**: It is the rejection of the null hypothesis H_o when it is true. The probability of this error is called α.
- **Type II**: It is the acceptance of the null hypothesis H_o when it is false. The probability of this error is called β. The term "*Power*" is used and it is defined as $1 - \beta$, the probability of rejecting the null hypothesis H_o when H_o is false.

The probabilities of Type I and Type II errors are inversely related. An increase in α results in decreased β, and vice versa. Typically, you will find α values ranging from 0.01 to 0.10 across the literature, representing the confidence level to reject/accept the hypothesis. In the end, it is the decision of the scientist what level of confidence will be used to assess the experimental results by statistical methods. Commonly, the hypotheses are tested considering both types of errors.

d. **Statistical Tests: Mean (μ) – Comparison of Treatments**

The following statistical tests are commonly used to assess experimental results (e.g., established hypotheses). Additional details associated with these tests are discussed elsewhere (e.g., Steel and Torrie, 1980; Montgomery, 1984; Ott, 1993). Also, these tests are codified or pre-programmed in commercial software available to the scientific community (e.g., EXCEL®, MINITAB®, SAS®, JUMP®, etc).

Our goal is to provide the readers with a high-level summary of the tests and how to use them.

- ***Statistical Test with known μ and σ (adapted from Ott, 1993; Montgomery, 1984).***

In this case, the test statistic uses the **z** distribution as shown in **Table 5.5.**

It is always recommended to assess the Type II error as part of the statistical test.

- ***Statistical Test with known μ and unknown σ (adapted from Ott, 1993; Montgomery, 1984).***
  In this case, the inferences on μ with an unknown σ require the use of the Student's *t* distribution developed by W.S. Gosset instead of the *z* distribution. The Student's *t* distribution approaches the *z* distribution when the number of samples is increased. There are different "*t*" distributions, the particular one to be considered in this chapter that introduces the parameter called *degrees of freedom*. Examples of the Student's *t* distribution are found elsewhere (Ott, 1993; Montgomery, 1984; Steel and Torrie, 1980; Kiemele et al., 2000). **Table 5.6** summarizes the statistical test with known μ and unknown σ.

TABLE 5.5 Statistical Test with Known μ *and* σ

Element	Definition	Comments
H_o	$\mu = \mu_o$	μ_o is specified
H_a	a) $\mu > \mu_o$ b) $\mu < \mu_o$ c) $\mu \neq \mu_o$	
Test	$z = \dfrac{\bar{y} - \mu_o}{\left(\dfrac{\sigma}{\sqrt{n}}\right)}$	If *n* is equal or greater than 30, σ can be replaced by the sample standard deviation *s* in the standard error.
Rejection region	For a probability α of Type I Error: a) Reject H_o if $z > z_\alpha$ b) Reject H_o if $z < -z_\alpha$ c) Reject H_o if $\lvert z \rvert > z_{\alpha/2}$	z_α and $z_{\alpha/2}$ are values from the cumulative standard normal distribution table.

TABLE 5.6 Statistical Test with Known μ and Unknown σ

Element	Definition	Comments
H_o	$\mu = \mu_o$	μ_o is specified
H_a	a) $\mu > \mu_o$ b) $\mu < \mu_o$ c) $\mu \neq \mu_o$	
Test	$t = \dfrac{\bar{y} - \mu_o}{\left(\dfrac{s}{\sqrt{n}}\right)}$	Since σ is unknown, it is replaced by the sample standard deviation s in the standard error.
Rejection region	For a probability α of Type I Error and degrees of freedom (df) = n−1: a) Reject H_o if $t > t_\alpha$ b) Reject H_o if $t < -t_\alpha$ c) Reject H_o if $\|t\| > t_{\alpha/2}$	t_α and $t_{\alpha/2}$ are values from the Student's t table, which are also a function of the degrees of freedom (df). Also, the tables for the Student's t distribution defines the term "a" which is same as α for one-tailed tests or $\alpha/2$ for two tails test.

- ***Statistical Test for comparison of $\mu_1 - \mu_2$: (1) independent samples with similar variance and (2) independent samples with different variance (adapted from Ott, 1993; Montgomery, 1984; Steel and Torrie, 1980).***

  It is a common situation to compare two sets of samples from independent populations during experimental work (e.g., control run vs. study run). These comparisons use the difference in the mean values ($\mu_1 - \mu_2$), where 1 and 2 represent each of the populations under evaluation. The following assumptions are considered: (1 independent samples; (2) normality; and (3) equal variances. In the situation where normality and equal variances are not valid, but sample sizes are large, the use of the t test is accepted (Ott, 1993).

Table 5.7 summarizes the hypotheses considering the case of similar variance on both independent samples.

Meanwhile, **Table 5.8** summarizes the hypotheses considering the case of different variance on both independent samples:

TABLE 5.7 Statistical Test for Comparison of $\mu_1 - \mu_2$: Independent Samples with Similar Variance

Element	Definition	Comments
H_o	$\mu_1 - \mu_2 = D_0$	D_o is specified
H_a	a) $\mu_1 - \mu_2 > D_0$ b) $\mu_1 - \mu_2 < D_0$ c) $\mu_1 - \mu_2 \neq D_0$	
Test	$t = \frac{(\bar{y}_1 - \bar{y}_2 - D_0)}{\left(s_p\sqrt{\frac{1}{n_1} + \frac{1}{n_2}}\right)}$ $s_p = \sqrt{\frac{(n_1 - 1)s_1^2 + (n_2 - 1)s_2^2}{n_1 + n_2 - 2}}$ $df = n_1 + n_2 - 2$	Where: $\bar{y}_i$ = mean value per population n_i = number of samples s_p = weighted average of standard deviations df = degrees of freedom
Rejection region	For a probability α of Type I Error and degrees of freedom (df): a) Reject H_o if $t > t_\alpha$ b) Reject H_o if $t < -t_\alpha$ c) Reject H_o if $\lvert t \rvert > t_{\alpha/2}$	t_α and $t_{\alpha/2}$ are values from the Student's *t* table, which are also a function of the degrees of freedom (df). Also, the tables for the Student's *t* distribution define the term "a" which is the same as α for one-tailed tests or $\alpha/2$ for two tails test.

- ***Nonparametric Alternative to Compare two Populations – Wilcoxon Rank Sum Test (adapted from Ott, 1993; Steel and Torrie, 1980).***
  A nonparametric alternative for comparing two populations is the Wilcoxon Rank Sum Test (Ott, 1993; Steel and Torrie, 1980). This test requires only that the two samples be independent and randomly drawn from their respective populations.
  The procedure involves ranking all observations from both populations together, assigning a rank of 1 to the smallest observation, 2 to the next smallest, and so on.
  When two or more observations have the same value, they are assigned the average of the ranks they would otherwise occupy. For example, if observations corresponding to ranks 4, 5, and 6 are equal, each is assigned a rank of 5 $[(4+5+6)/3]$. Similarly, if observations at ranks 7 and 8 are equal, each receives a rank of 7.5 $[(7+8)/2]$. After ranking, the number of tied observations at each rank is recorded, as only tied ranks affect the variance of the test statistic.

TABLE 5.8 Statistical Test for Comparison of $\mu_1 - \mu_2$: Independent Samples with Different Variance

Element	Definition	Comments
H_o	$\mu_1 - \mu_2 = D_0$	D_o is specified
H_a	a) $\mu_1 - \mu_2 > D_0$ b) $\mu_1 - \mu_2 < D_0$ c) $\mu_1 - \mu_2 \neq D_0$	
Test	$t' = \frac{(\bar{y}_1 - \bar{y}_2 - D_0)}{\left(\sqrt{\frac{s_1^2}{n_1} + \frac{s_2^2}{n_2}}\right)}$ $df = \frac{(n_1-1)(n_2-1)}{(n_2-1)c^2 + (1-c)^2(n_1-1)}$ $c = \frac{\left(s_1^2 / n_1\right)}{\left(s_1^2 / n_1 + s_2^2 / n_2\right)}$	Where: t' = an approximate t test $\bar{y}_i$ = mean value per population n_i = number of samples s^2_i = variance value per population df = degrees of freedom, rounded down to the nearest integer.
Rejection region	For a probability α of Type I Error and degrees of freedom (df): a) Reject H_o if $t' > t_\alpha$ b) Reject H_o if $t' < -t_\alpha$ c) Reject H_o if $\lvert t' \rvert > t_{\alpha/2}$	t_α and $t_{\alpha/2}$ are values from the Student's t table, which are also a function of the degrees of freedom (df). Also, the tables for the Student's t distribution define the term "a" which is the same as α for one-tailed tests or $\alpha/2$ for two tails test.

For sample sizes greater than 10, the Wilcoxon Rank Sum Test statistic is approximated using the standard normal (z) distribution and is calculated using the following formula.

$$z = \frac{T - \mu_T}{\sigma_T}$$

where

- T is the sum of the ranks in sample 1 from Population 1.
- $\mu_T = n_1 * (n_1 + n_2 + 1) / 2$
- n_1 and n_2 are the number of observations in each experiment.
- $\sigma_T^2 = \left(\frac{n_1 n_2}{12}\right)\left[(n_1 + n_2 + 1) - \frac{\sum_i t_i (t_i^2 - 1)}{(n_1 + n_2)(n_1 + n_2 - 1)}\right]$

- t_i represents the number of tied (repeated) ranks in the ith group. If no tied ranks are observed, then the variance is:

$$\sigma_T^2 = \left(\frac{n_1 n_2}{12}\right)\left[n_1 + n_2 + 1\right]$$

The Wilcoxon Rank Sum Test is more informative when the assumptions for the t test are violated, especially if non-normality of the populations is present (e.g., extreme outliers or skewness). In summary, the hypotheses to be evaluated using the Wilcoxon Rank Sum Test are summarized in **Table 5.9.**

TABLE 5.9 Hypotheses Test – Wilcoxon Rank Sum

Element	**Definition**	**Comments**
H_o	Two populations are identical	
H_a	a) Population 1 is shifted to right of Population 2. b) Population 1 is shifted to left of Population 2. c) Populations 1 and 2 have different location parameters.	Populations may have non-normal distributions.
Test	***For n_1 and $n_2 \leq 10$:*** T, the sum of the ranks in sample 1 ***For n_1 and $n_2 > 10$:*** $z = \frac{T - \mu_T}{\sigma_T}$	Nonparametric calculations to compare the populations focus on "ranks."
Rejection region	For a probability α of Type I Error: ***For n_1 and $n_2 \leq 10$:*** Use the Critical Values of T_L and T_U for the Wilcoxon Rank Sum Test published elsewhere (Ott, 1993): a) Reject H_o if $T > T_U$ b) Reject H_o if $T < T_L$ c) Reject H_o if $T > T_U$ or $T < T_L$ ***For n_1 and $n_2 > 10$:*** Use the cumulative standard normal distribution table: a) Reject H_o if $z > z_\alpha$ b) Reject H_o if $z < -z_\alpha$ c) Reject H_o if $\lvert z \rvert > z_{\alpha/2}$	

- ***Paired t Test (adapted from Ott, 1993; Montgomery, 1984).*** Experiments where measurements in a sample are paired or matched with a second measurement under a second sample (e.g., cost of repairs from two shops for the same snowmobile). In this particular case, the samples are dependent and require the use of a different analysis than those previously discussed.
  The Paired t Test is based on the actual difference (d) between the n paired samples:
  $d_i = (\mu_1 - \mu_2)I$ (individual difference)

  $$\bar{d} = \frac{\sum_i d_i}{n} \quad \text{(mean of the differences)}$$

  $$s_d^2 = \frac{1}{n-1}\left[\sum_i d_i^2 - \frac{\left(\sum_i d_i\right)^2}{n}\right] \quad \text{(variance of the n differences)}$$

  where i represent each of the n samples under evaluation. The statistics are summarized in **Table 5.10**.

TABLE 5.10 Paired t Test

Element	Definition	Comments
H_o	$\mu_d = D_o$	D_o is the specified target difference.
H_a	a) $\mu_d > D_o$ b) $\mu_d < D_o$ c) $\mu_d \neq D_o$	
Test	$t = \frac{\bar{d} - D_o}{\left(\frac{s_d}{\sqrt{n}}\right)}$ where $\bar{d}$ and s_d are the sample mean and standard deviation of the n differences.	All calculations are focused on the n differences.
Rejection region	For a probability α of Type I Error and degrees of freedom (df) = $n-1$: a) Reject H_o if $t > t_\alpha$ b) Reject H_o if $t < -t_\alpha$ c) Reject H_o if $\lvert t \rvert > t_{\alpha/2}$	t_α and $t_{\alpha/2}$ are values from the Student's t table, which are also a function of the degrees of freedom (df). Also, the tables for the Student's t distribution defines the term "a" which is same as α for one-tailed tests or $\alpha/2$ for two tails test.

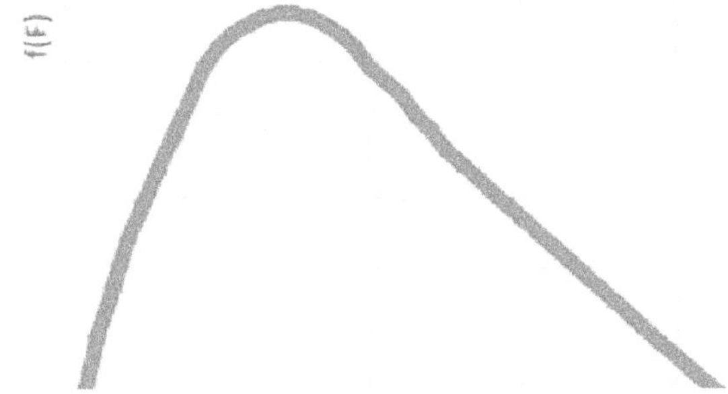

FIGURE 5.9 Representation of *F* distribution.

- ***Comparison of Variances using the F distribution (adapted from Ott, 1993; Montgomery, 1984).***

  This test is used for checking the validity of equal variances during the *t* test of two samples. Independent random samples will have a probability distribution of the ratio of their variances represented by the *F* distribution **(Figure 5.9)**.

$$Ratio\ of\ variances = \frac{s_1^2 / s_2^2}{\sigma_1^2 / \sigma_2^2}$$

The *F* distribution has the following characteristics:

- The Degrees of Freedom (e.g., df_1 and df_2) associated with each variance (e.g., $s_1^2 and s_2^2$) are used to specify an *F* distribution as there are many *F* distributions.
- The *F* only assumes positive values.
- It is non-symmetrical.
- Tail values for the *F* distribution are published elsewhere (Ott, 1993; Montgomery, 1984; Steel and Torrie, 1980; Kiemele et al., 2000).

Table 5.11 presents a summary of the test procedure.

These principles will be used when discussing the concepts of Analysis of Variance for One-Way and Multiway cases.

e. **Analysis of Variance (ANOVA)**

The Analysis of Variance was initially introduced by Sir Ronald A. Fisher in 1918 in his publication *The Correlation between Relatives on the Supposition of Mendelian Inheritance* followed by the publication

TABLE 5.11 Comparison of Variances Using the *F* Distribution

Element	Definition	Comments
H_o	$\sigma_1^2 = \sigma_2^2$	
H_a	a) $\sigma_1^2 > \sigma_2^2$ b) $\sigma_1^2 \neq \sigma_2^2$	
Test	$F = \frac{s_1^2}{s_2^2}$	
Rejection region	For a probability α of Type I Error and degrees of freedom, $df_i = n_i - 1$: a) Reject H_o if $F > F_{a,df_1,df_2}$ for a = α b) Reject H_o if $F > F_{a,df_1,df_2}$ for $a = \alpha/2$	F_{a,df_1,df_2} are found tabulated elsewhere (e.g., Ott, 1993; Montgomery, 1984; Steel and Torrie, 1980; Kiemele et al., 2000)

of the article "Statistical Tests of Agreement between Observation and Hypothesis" in 1923, where he formalized significance testing and connected variance ratios to probability distributions. The analysis is basically the partitioning of the total sum of squares into components associated with recognized sources of variation (Steel and Torrie, 1980). The basic assumptions associated with ANOVA are the following:

- Experimental errors are random, independent, and normally distributed around a zero mean and with a common variance.
- Treatment and environmental effects are additive.

This section of the chapter focuses on three specific cases:

- One Way ANOVA.
- Multiway ANOVA.
- Factorial Experiments.

- ***ANOVA – One-Way (t number of treatments with r number of replicates)***

  The One-Way ANOVA allows the comparison of multiple treatments or experiments (*t*) simultaneously. Experimental units associated with the experiments are homogeneous or have relatively small differences among them. Also, common experiments include multiple replicates of the experimental units. The procedure for the ANOVA is summarized in **Table 5.12**.

TABLE 5.12 ANOVA - One-Way

Source of Variance	Degrees of Freedom (df)	Sum of Squares	Mean Squares	F Value
Treatments	$df_t = t - 1$	$SS_t = r\sum_i \left(\overline{Y}_{i.} - \overline{Y}_{..}\right)^2$	$MS_t = \frac{SS_t}{(t-1)}$	$F = \frac{MS_t}{MS_e}$
Error	$df_e = t(r - 1)$	$SS_e = \sum_{i,j} \left(Y_{ij} - \overline{Y}_{i.}\right)^2$	$MS_e = \frac{SS_e}{t(r-1)}$	
Totals	$rt - 1$	$\sum_{i,j} \left(Y_{ij} - \overline{Y}_{..}\right)^2$		

where:

t is the number of treatments.
r is the number of replicates.
Y_{ij} are each of the individual observations (i = treatment, j = replicate).
$\overline{Y}_{i.}$ are the average per treatment $[(\sum_j Y_{ij})/r]$.

$\overline{Y}_{..}$ is the overall average $[(\sum_i \overline{Y}_{i.})/t$ or $\left(\sum_{i,j} Y_{ij}\right)/(rt)]$.

The hypothesis testing is summarized in **Table 5.13**.

There are additional test options considering the One-Way ANOVA such as: (1) ANOVA with unequal replication and (2) ANOVA with subsamples – equal and unequal subsample numbers. Additional information can be found elsewhere.

TABLE 5.13 Hypothesis Testing Using ANOVA-One Way

Element	Definition	Comments
H_o	$\overline{Y}_{1.} = \overline{Y}_{2.} \ldots\ldots = \overline{Y}_{t.}$	
H_a	There are real differences among treatment means	
Test	$F = \frac{MS_t}{MS_e}$	
Rejection region	For a probability α of Type I Error and degrees of freedom, $df_t = t - 1$ and $df_e = t(r - 1)$: Reject H_o if $F > F_{a,df_t,df_e}$	F_{a,df_t,df_e} are found tabulated elsewhere (e.g., Ott, 1993; Montgomery, 1984; Steel and Torrie, 1980; Kiemele et al., 2000).

- ***ANOVA – Multiway***

  Completely random designs are appropriate when the only source of variation is the treatment's effects. In certain cases, it is known ahead of the experiments that some experimental units will behave differently, even when treated under the same conditions. Factors such as temperature fluctuations, air velocity, sunlight, time of day, or history of fertilizers on soils will introduce variability that could affect the experimental results. Therefore, proper design of experiments will ensure such variability is accounted for and removed from the experimental error.

  The use of Blocks or Latin Squares designs allows the Analysis of Variance where two or more factors are under evaluation.

  - *Randomized Complete Block Design*

    The design requires meaningful grouping of the experimental units to block a particular source of variability. The number of treatments may or may not be used to define the number of blocks. The goal of grouping is to have the units in a block as uniform as possible to ensure that observed differences are largely caused by treatments. During the experimental phase, all units in a block must be treated as uniformly as possible except for the treatment. Any change in handling affecting a unit in a block must be applied to every other unit in that block.

  Randomization is conducted within the block by random assignment of the treatments to the units in a block. The randomization is independent in the block. The procedure for the ANOVA is summarized in **Table 5.14**.

TABLE 5.14 ANOVA – Randomized Complete Block Design

Source of Variance	Degrees of Freedom (df)	Sum of Squares	Mean Squares	F Value
Blocks	$df_B = r - 1$	$SS_B = t\sum_j \left(\bar{Y}_{.j} - \bar{Y}_{..}\right)^2$	$MS_B = \frac{SS_B}{(r-1)}$	
Treatments	$df_{Treat} = t - 1$	$SS_{Treat} = r\sum_i \left(\bar{Y}_{i.} - \bar{Y}_{..}\right)^2$	$MS_{Treat} = \frac{SS_{Treat}}{(t-1)}$	$F = \frac{MS_{Treat}}{MS_e}$
Error	$df_e = (t-1)(r-1)$	$SS_e = \sum_{i,j}\left(Y_{ij} - \bar{Y}_{.j} - \bar{Y}_{i.} + \bar{Y}_{..}\right)^2$	$MS_e = \frac{SS_e}{(t-1)(r-1)}$	
Totals	$rt - 1$	$\sum_{i,j}\left(Y_{ij} - \bar{Y}_{..}\right)^2$		

where:

t is the number of treatments.
r is the number of blocks.
Y_{ij} are each of the individual observations (i = treatment, j = replicate).
$\overline{Y}_{i.}$ are the averages per treatment $[\left(\sum_j Y_{ij}\right)/r]$.
$\overline{Y}_{.j}$ are the average per block $[\left(\sum_i Y_{ij}\right)/t]$.
$\overline{Y}_{..}$ is the overall average $[\left(\sum_{i,j} Y_{ij}\right)/(rt)]$.
SS = Sum of Squares.
MS = Mean Squares.

The hypothesis testing is summarized in **Table 5.15**.

- *Latin Squares*
  The Latin Square is a case where treatments are arranged in blocks in two different ways: rows and columns. Each treatment will occur once in each row and column. The design is used when two major sources of variation, besides the treatments, are present during the execution of the experiments.
  The rows and columns may be considered some kind of treatment. If there is any interaction among the columns, rows, and treatments, then the analysis is not valid (Steel and Torrie, 1980). Therefore, if

TABLE 5.15 Hypothesis Testing Using Randomized Complete Block Design

Element	**Definition**	**Comments**
H_o	$\overline{Y}_{1.} = \overline{Y}_{2.} \ldots\ldots = \overline{Y}_{t.}$	
H_a	There are real differences among treatment means	
Test	$F = \dfrac{MS_{Treat}}{MS_e}$	
Rejection region	For a probability α of Type I Error and degrees of freedom, $df_{Treat} = t - 1$ and $df_e = (t-1)(r-1)$: Reject H_o if $F > F_{a,df_{Treat},df_e}$	F_{a,df_{Treat},df_e} are found tabulated elsewhere (e.g., Ott, 1993; Montgomery, 1984; Steel and Torrie, 1980; Kiemele et al., 2000).

the scientist is not able to rule out the presence of interactions, this design should not be used.

The Latin Square Design requires an equal number of rows, columns, and treatments, which represents a main disadvantage. The most common design is in the range of 5 × 5 to 8 × 8. However, a square may be replicated to increase the number of degrees of freedom (e.g., a single 4 × 4 square will have 15 degrees of freedom, while replicating the square will result in 31 degrees of freedom).

Randomization in the square consists of choosing a square at random from all possible options. The procedure for the ANOVA is summarized in **Table 5.16.**

TABLE 5.16 ANOVA – Latin Square Design

Source of Variance	Degrees of Freedom (df)	Sum of Squares	Mean Squares	F Value
Columns	$df_C = r - 1$	$SS_C = r\sum_j \left(\bar{Y}_{.j} - \bar{Y}_{..}\right)^2$	$MS_C = \frac{SS_C}{(r-1)}$	$F = \frac{MS_C}{MS_e}$
Rows	$df_R = r - 1$	$SS_R = r\sum_i \left(\bar{Y}_{i.} - \bar{Y}_{..}\right)^2$	$MS_R = \frac{SS_R}{(r-1)}$	$F = \frac{MS_R}{MS_e}$
Treatments	$df_{Treat} = r - 1$	$SS_{Treat} = r\sum_t \left(\bar{Y}_t - \bar{Y}_{..}\right)^2$	$MS_{Treat} = \frac{SS_{Treat}}{(r-1)}$	$F = \frac{MS_{Treat}}{MS_e}$
Error	$df_e = (r - 1)(r - 2)$	$SS_e = \sum_{i,j} \left(Y_{ij} - \bar{Y}_{.j} - \bar{Y}_{i.} - \bar{Y}_t + 2\bar{Y}_{..}\right)^2$	$MS_e = \frac{SS_e}{(r-1)(r-2)}$	
Totals	$r^2 - 1$	$\sum_{i,j} \left(Y_{ij} - \bar{Y}_{..}\right)^2$		

where:

r is the number of rows, columns, and treatments.

Y_{ij} are each of the individual observations (i = rows, j = columns).

$\bar{Y}_{i.}$ are the averages per row $[\left(\sum_j Y_{ij}\right)/r]$.

$\bar{Y}_{.j}$ are the average per column $[\left(\sum_i Y_{ij}\right)/r]$.

$\bar{Y}_t$ is the average for the treatments $[\left(Y_1 + Y_2 + \ldots Y_r\right)/r]$.

$\bar{Y}_{..}$ is the overall average $\left(\sum_{i,j} Y_{ij}\right)/\left(r^2\right)]$.

SS = Sum of Squares.
MS = Mean Squares.

The hypothesis testing is summarized in **Table 5.17**.

TABLE 5.17 Hypothesis Testing Using Latin Square

Element	Definition	Comments
H_o	$Y_1 = Y_2 = \ldots\ldots Y_r$	This can be extended to the rows and columns.
H_a	There are real differences between the treatments.	This can be used for rows and columns.
Test	$F = \frac{MS_{Treat}}{MS_e}$	This can be used for rows and columns.
Rejection region	For a probability α of Type I Error and degrees of freedom, $df_{Treat} = r - 1$ and $df_e = (r - 1)(r - 2)$: Reject H_o if $F > F_{a,dfTreat,dfe}$	$F_{a,dfTreat,dfe}$ are found tabulated elsewhere (e.g., Ott, 1993; Montgomery, 1984; Steel and Torrie, 1980; Kiemele et al., 2000).

- ***Factorial Experiments***
  In Factorial Experiments there are a number of treatments in each of multiple categories which define a treatment grid (Steel and Torrie, 1980.) A factor is defined as a kind of treatment which may have multiple levels (e.g., if sun exposure time is a factor, then multiple times (levels) of sun exposure will be considered in the experiment; if the concentration of NaCl is a factor, then multiple NaCl concentrations (levels) are considered).
  These experiments are common across all fields of research. They provide great value in experimental work where there is little knowledge (e.g., optimum levels or relevance) of the factors under study. Among the advantages of Factorial Designs are the following: (1) the flexibility to assess possible interactions between the factors under evaluation and (2) allowing effects of a factor to be estimated at multiple levels of the other factors. All conclusions will be valid over the range of experimental conditions (Montgomery, 1984).
  The ANOVA for this type of experiments will consider each factor as well as the interactions. The procedure for the ANOVA considering a 2-Factor Factorial Experiment is summarized in **Table 5.18**.

TABLE 5.18 ANOVA – Two-Factors Factorial Experiment

Source of Variance	Degrees of Freedom (df)	Sum of Squares	Mean Squares	F Value
A Treatments	$df_A = a - 1$	$SS_A = \sum_{i=1}^{a} \frac{y_{i..}^2}{bn} - \frac{y_{...}^2}{abn}$	$MS_A = \frac{SS_A}{(a-1)}$	$F = \frac{MS_A}{MS_e}$
B Treatments	$df_B = b - 1$	$SS_B = \sum_{j=1}^{b} \frac{y_{.j.}^2}{an} - \frac{y_{...}^2}{abn}$	$MS_B = \frac{SS_B}{(b-1)}$	$F = \frac{MS_B}{MS_e}$
		$SS_{Subtotals} - \sum_{i-1}^{a}\sum_{j=1}^{b} \frac{y_{ij.}^2}{n} - \frac{y_{...}^2}{abn}$		
Interactions	$df_{Int} = (a - 1)(b - 1)$	$SS_{Int} = SS_{Subtotals} - SS_A - SS_B$	$MS_{Int} = \frac{SS_{Int}}{(a-1)(b-1)}$	$F = \frac{MS_{Int}}{MS_e}$
Error	$df_e = ab(n - 1)$	$SS_e = SS_T - SS_{Int} - SS_A - SS_B$	$MS_e = \frac{SS_e}{ab(n-1)}$	
Totals	$df_T = abn - 1$	$SS_T = \sum_{i=1}^{a}\sum_{j-1}^{b}\sum_{k=1}^{n} y_{ijk}^2 - \frac{y_{...}^2}{abn}$		

Where:

a is the number of treatments A.

b is the number of treatments B.

n is the total of observations per experiment.

y_{ijk} are each of the individual observations (i = Treatment A, j = Treatment B, k = Observation).

$$y_{i..} = \sum_{j=1}^{b}\sum_{k=1}^{n} y_{ijk} \qquad \bar{y}_{i..} = \frac{y_{i..}}{bn} \qquad i = 1,2,3....a$$

$$y_{.j.} = \sum_{i=1}^{a}\sum_{k=1}^{n} y_{ijk} \qquad \bar{y}_{.j.} = \frac{y_{.j.}}{an} \qquad j = 1,2,3....b$$

$$y_{ij.} = \sum_{k=1}^{n} y_{ijk} \qquad \bar{y}_{ij.} = \frac{y_{ij.}}{n} \qquad i = 1,2,....a \text{ and } j = 1,2,.....b$$

$$y_{...} = \sum_{i=1}^{a}\sum_{j=1}^{b}\sum_{k=1}^{n} y_{ijk} \qquad \bar{y}_{...} = \frac{y_{...}}{abn}$$

SS = Sum of Squares
MS = Mean Squares

The hypothesis testing is summarized in **Table 5.19**.

TABLE 5.19 Hypothesis Testing – Two-Factor Factorial Experiment

Element	Definition	Comments
H_o	There is no interaction between the main factors. Factor A and/or B have no effect.	
H_a	There is a significant interaction between the main factors. Factor A and/or B have a significant effect.	
Test	$F = \frac{MS_A}{MS_e}$ $F = \frac{MS_B}{MS_e}$ $F = \frac{MS_{Int}}{MS_e}$	
Rejection region	For a probability α of Type I Error and degrees of freedom, $df_A = a - 1$, $df_B = b - 1$, $df_{Int} = (a - 1)(b - 1)$, and $df_e = ab(n - 1)$: Reject H_o if $F > F_{a, dfA,B, or\ Int, dfe}$	$F_{a, dfA,B\ or\ Int, dfe}$ are found tabulated elsewhere (e.g., Ott, 1993; Montgomery, 1984; Steel and Torrie, 1980; Kiemele et al., 2000).

If additional factors are considered for an experiment (e.g., 3-Factor Factorial), additional adjustments are made to the equations and calculations previously discussed. The interactions between factors will now consider AB, AC, BC, and ABC in a 3-Factor Factorial Design. In some cases, the scientist may decide on "confounding" some of these elements. Confounding is a technique for arranging a factorial experiment in blocks. This technique causes information about certain effects to be mixed with blocks and not able to distinguish individual effects. In other words, an interaction will be part of the result for a particular factor, and there is no way to separate the two. In several instances, the results of an interaction are not significant compared to the actual result from a single factor.

5.4 CONCLUSION FROM EXPERIMENTS

The next step after the data analysis is to reach a conclusion. Using as a reference the problem statement, hypotheses, and the statistical analyses, the scientist should be able to provide a conclusion. The following steps are recommended as part of the experimental work:

- Verify all the raw data and confirm that the data used for statistical analyses are correct.
- Confirm that the test methods were properly executed and satisfied the requirements of the study.

- Confirm that the statistical analysis is aligned with the design of experiments and that data was collected according to the design (e.g., randomization)
- When using a software application, ensure that the variables are properly defined, as well as any calculations within the software.
- Check for any atypical results and investigate. If a clear root cause is identified (e.g., error on execution, contamination, wrong components), invalidate the results and proceed with re-execution of the study.
- Trust science! Accept the results and the analysis of the data. Conclusions must be based on facts.

Artificial Intelligence (AI) applications (e.g., ChatGPT®, Claude, MS Copilot®, Google Gemini, Notion AI, etc.) are increasingly assisting scientists and research groups in analyzing data and formulating conclusions. Nevertheless, as with any technological tool, effective use of AI requires appropriate data governance and careful contextual interpretation to ensure reliable and valid results.

5.5 FINAL COMMENTS

The conclusions from any experimental study are based on the data and associated analysis. The inferences must be limited to the study constraints and compiled data. The statistical tools discussed in this chapter, as well as others available elsewhere, are key resources that facilitate reaching conclusions. It is up to the scientist to properly use the tools as part of any experimental work.

5.6 REFERENCES

Fisher, R.A. 1918. The Correlation between Relatives on the Supposition of Mendelian Inheritance. *Transactions of the Royal Society of Edinburgh* 52(2): 399–433. https://doi.org/10.1017/S0080456800012163.

Fisher, R.A. 1923. Statistical Tests of Agreement Between Observation and Hypothesis. *Economica* (8): 139–147. https://doi.org/10.2307/2548482.

Kiemele, M.J., Schmidt, S.R., and Berdine, R.J. 2000. *Basic Statistics: Tools for Continuous Improvement.* 4th Edition. Air Academy Press, LLC, Colorado Spring, CO.

Montgomery, D.C. 1984. *Design and Analysis of Experiments.* 2nd Edition. Joh Wiley & Sons, Inc., New York, NY.

Ott, L. 1993. *An Introduction to Statistical Methods and Data Analysis.* 4th Edition. Duxbury Press, Belmont, CA.

Steel, R.G.D., and Torrie, J.H. 1980. *Principle and Procedures of Statistics: A Biometrical Approach.* 2nd Edition. McGraw Hill, Inc., New York, NY.

CHAPTER 6

Reports and Presentations

6.1 INTRODUCTION

The primary purpose of reporting, presenting, and publishing research is to communicate findings to the scientific community. This communication should be clear, concise, and adhere to standards established by relevant scientific organizations within each discipline.

This chapter provides the readers with an overview of the techniques and styles commonly used in scientific reports, presentations, and publications at local, national, and international organizations and forums.

6.2 GLOSSARY OF TERMS

The following terms, commonly used by scientists, are used in this chapter.

Term	Definition
Abstract	Short summary that explains purpose, methods, results, and conclusion.
Balance on slides	Even distribution of text and images on slides.
Conclusion	The final section of a paper or presentation that summarizes the main points and explains the overall meaning or importance of the work.
Cover page	First page of a report that includes title, authors, organization, and dates

DOI: 10.1201/9781042010516-6

Term	Definition
Discussion	Written explanation of existing information from literature or observed results during an experiment.
Letter fonts	Style and appearance of text on a written document.
Materials	Compounds and components used in a project or experiment.
Memorandum	Short and formal message used within a group.
Methods	Procedures or steps required to perform an activity, experiment, or test.
Oral presentation	A spoken explanation of information to an audience, often supported by slides or visual aids.
Plagiarism	The act of using someone else's words, ideas, or work without giving proper credit.
Poster	A visual presentation that summarizes research or information containing text, plots, diagrams, and pictures.
Publication	Digital or printed material made available to the public.
Recognitions	Acknowledgement of people, organizations, or sponsors supporting the work.
References	Comprehensive list of all sources (e.g., articles, books, and publications) used while researching a topic. It gives credit and avoids plagiarism.
Reports	Written formal document presenting information clearly and logically about a subject.
Results	Outcome of an activity or experiment.
Table of contents	Comprehensive list of the contents with page numbers (e.g., report, book, journal).
Third rule	Center information in a 3 × 3 grid; also refer to "Rule of Three," to limit information to no more than three main points.
Visual format	Layout and design of a document or presentation.

6.3 REPORTS

Periodic and final reports are utilized to keep teachers, colleagues, associates, teammates, and project leaders up to speed about progress, challenges, and next steps in a project. The reports often serve as foundational work for eventual presentations and journal publications. In addition, these periodic and final reports identify areas of future research on the topic at hand.

The format of the reports should be standardized for easy compilation of historical information. Also, the use of best practices observed in the scientific community is recommended. Those practices include, but are not limited to, the use of Good Laboratory Practices (GLP) and Good Documentation Practices (GDocP) promoted by regulatory bodies and professional organizations such as the US Department of Agriculture, the US Food and Drug Administration, the European Medicines Agency, the Australian Health Authority (TGA), the Parenteral Drug Association (PDA), and the International Society of Pharmaceutical Engineers (ISPE).

- *Progress Reports or Partial Reports*
  A progress report or partial report focuses on work or results at a particular point in time or with a particular period, step, or stage during the lifecycle of a project. A common practice is to issue weekly, biweekly, or monthly reports focused on specific aspects of the scientific project. The basic elements of periodic reports are as follows:
  - *Their combination serves as a basis for presentations at conferences or publications.*
  - *Close monitoring of the progress in experimental work and data analysis.*
  - *Discussion of potential issues and suggestions involving re-definition of activities or revisiting hypotheses to be further considered by the investigation team.*

  Reports should be brief and concise while including summaries of experimental data either in tables or plots. This allows for periodic peer review of the work and facilitates early detection of errors and issues by those receiving the reports.
  An example of the structure and content of a partial report is presented in **Figure 6.1.**
- *Final Reports*
  The final report consolidates all information developed over the entire lifecycle of the project, from the problem statement through the conclusions. The title of this report varies depending on context, for example: (1) it is referred to as a *Thesis* or *Dissertation* when produced by undergraduate, master's, or PhD students as part of degree requirements; (2) as a *White Paper* when prepared by scientists or engineers in a corporate setting; and (3) as a *Technical Report* when authored by professionals within a professional organization (e.g., IFT, PDA, ISPE, etc.). Regardless of its designation, the final report is a comprehensive document that provides detailed coverage of the

Memorandum

To: Group Leads, Managers, etc.
From: Lead Investigator, Co-Authors
Subject: Progress Report Project ABC – Date1 to Date 2
Date: DD-MMM-YYYY

The following activities have been conducted as part of Project ABC:

1. Purpose and Hypotheses
   State the goal of the work and the hypotheses under evaluation
2. Research Activities:
   Summary of work conducted by the team.
3. Design of Experiments:
   Summary of progress involving DoE.
4. Experiments and Results:
   Summary of progress on experimental work and results.
5. Presentations and Publications
   Summary of progress on presentations and publications
6. Issues, Challenges and Opportunities:
   Summary of issues encountered, challenges affecting the project, and opportunities identified as part of the project.

FIGURE 6.1 Format of a partial/progress report.

subject matter. Careful attention must be given to page numbering and its accurate linkage to the Table of Contents to facilitate efficient navigation for readers. The final format and structure of the report should be discussed and approved by the research team and/or the graduate committee. **Table 6.1** summarizes the contents of final reports.

6.4 PRESENTATIONS

Presentations are one of several mechanisms used to communicate the progress and outcomes of research activities. Events such as symposia, conferences, technical talks, scientific fairs, and workshops provide platforms for sharing the latest developments in research and development within the scientific community. Scientists typically use either poster or oral presentations to disseminate their work. Poster presentations emphasize visual elements – such as prominent titles, images, plots, and tables – to attract and engage audiences, whereas oral presentations focus on the presenter's delivery, expertise, and ability to effectively communicate the subject matter. **Table 6.2** presents a list of commercially available software

TABLE 6.1 Final Reports – Overview of Content

Section	Content
Cover page	Title, Organization, Dates, and Authors.
Table of contents	Title of chapter headings and sections with the assigned pages.
Introduction	Comprehensive discussion of the topic under investigation, goal of the project, potential hypotheses, and justification to proceed with the project.
Literature review	Summary of existing information on the subject matter.
Material and methods	Description of the procedures and analytical methods used.
Results and discussion	Comprehensive discussion of the test results, raw data, and plots showing the outcomes from all experimental work.
Conclusions	Consolidation of results and outcome of hypothesis testing.
References	List of references (e.g., articles, books, online sites, scientific communications with peers).

TABLE 6.2 Example of Software Available to Prepare Presentations

Software	Supplier/ Developer	Key Capabilities/Strengths
Microsoft PowerPoint	Microsoft	Industry standard slide creation; rich design tools, animations, templates, and deep Office integration. Works on desktops and the web.
Google Slides	Google LLC	Cloud-based real-time collaboration, revision history, easy sharing, integrates with Google Workspace.
Canva	Canva Pty Ltd.	Drag-and-drop visual design, huge template and asset library, brand kits, good for visually rich presentations.
Prezi	Prezi Inc.	Non-linear zoomable canvas storytelling, interactive movement, and spatial context, Prezi Video for online presentation.
Visme	Visme	Combines slide creation with data visualization and interactive content; good for branded, animated decks.
Pitch	Pitch (Pitch Software GmbH)	Modern web-based tools focused on team collaboration, version control, and sleek visuals.
Beautiful.ai	Beautiful.ai	AI-assisted layout and design automation so slides look professional with minimal manual formatting.
Slidebean	Slidebean	AI-driven slide design: automatically arranges content and designs based on best practices and templates.
Apple Keynote	Apple	Polished aesthetic themes, rich animations, and transitions; best for Mac/iOS ecosystem users.
Zoho Show	Zoho Corporation	Cloud-based editor with collaboration and broadcasting tools; integrates with the Zoho business suite.

that could be used to prepare presentations (e.g., posters and oral presentations). It is up to scientists to select and use the one that best suits their needs.

Basic considerations and recommendations are provided for the preparation of poster presentations and oral presentations.

- *Poster Presentations – Preparation and Display*
  The information for a poster is commonly arranged to fit a pre-defined surface area ranging from 20 to 24 square feet (e.g., 5' × 4' or 6' × 4'). The advances in digital technology allow one to print in a single poster all of the information, or the author(s) may consider printing each section of the poster on separate pages that are then attached to a poster board. **Table 6.3** presents the elements to be considered when preparing the content for a poster:

The following recommendations focus on the content of the poster [e.g., Introduction, Objectives, Materials and Methods, Results and Discussion (paragraphs, diagrams, plots, tables, and pictures), Conclusions and Recommendations, and Bibliography and Recognitions]:

1. Selected letter size and styles should allow reading of the content from a distance of about 5 feet. Use solid colors over white or clear colors, avoiding combinations that may strain the eyes and make it difficult to read the content. **Figure 6.2** presents some examples of letter sizes and styles (fonts) commonly used in posters.

TABLE 6.3 Elements to Consider While Preparing a Poster Presentation

Elements to Consider	Explanation
Title	A concise and meaningful title communicating the goal of the presentation. Size and color of the letters in the title should allow a person to read it while standing about 30–50 feet away.
Author(s) and organization	List the author's name(s) and the organization represented at the event below the title.
Content	Introduction, objectives, materials and methods, results and discussion (e.g., paragraphs, diagrams, plots, tables, and pictures), conclusions and recommendations, and bibliography and recognitions. The content shall be presented in a logical sequence. where all the components described above are blended to avoid overwhelming the reader.
Fill, outline, and effects	Frame and background colors, as well as shapes, should facilitate the review and discussion of the content.

2. Proofread the content to avoid errors (orthographic and typographic). Ensure uniform use of letter sizes and styles within the poster.
3. Use full short sentences while discussing a topic. Avoid long paragraphs.
4. Use pictures, plots, diagrams, and drawings as often as possible to summarize key ideas or outcomes. Limit the use of tables.

Once the content is finished, proceed to display the poster and carefully review it as a product of your work. Ask yourself a basic question: Am I satisfied with the structure and flow of the information? Also, ask your peers and friends for feedback. **Figure 6.3** shows an example of the arrangement for a poster presentation as well as a picture of a poster from a science fair competition.

As a final point, it is important to check with the sponsor of the event about the following details:

1. Where the poster is to be placed (e.g., wall, board, table).
2. How to attach the poster or keep it standing (e.g., tape, clips, glue, pins).

Legend: Letter – Size – Style		
Calibri – 11 – Normal	Arial – 11 – Normal	Time Roman – 11 – Normal
Calibri – 11 – Bold	**Arial – 11 – Bold**	**Time Roman – 11 – Bold**
Calibri – 11- Italics	*Arial – 11 - Italics*	*Time Roman – 11 – Italics*
Calibri – 18 – Normal		
Calibri – 18 – Bold		
Calibri – 18- Italics		
Time Roman – 36 – Normal		
Time Roman – 36 – Bold		
Time Roman – 36 – Italics		

FIGURE 6.2 Example of letter sizes and styles (fonts) useful on posters.

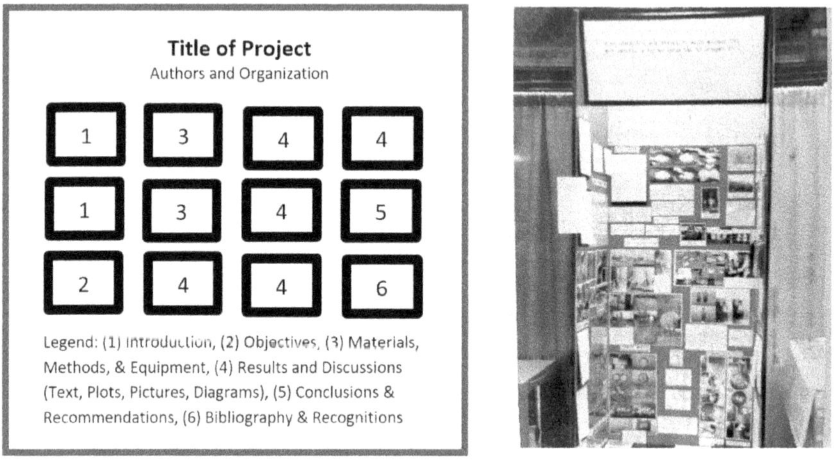

FIGURE 6.3 Example of the arrangement for a poster presentation as well as a picture of a poster from a science fair competition. *Source*: (H.Vega, 1980).

3. How long the poster will be exhibited, and when you should be available to answer questions about the poster.

- *Oral Presentations – Preparation and Discussion*
  The preparation of oral presentations follows a process similar to the development of poster presentations. The main difference between a poster and an oral presentation is not the "space" for the poster, but the allocated time for the presenter to discuss the content. Typical presentation times range from 15 to 30 minutes, followed by 5 to 10 minutes for questions and answers. The key elements for a successful oral presentation are summarized in **Table 6.4**.
  The slides for an oral presentation will be a combination of high-level information, in which the presenter provides additional details, and very pointed information (e.g., pictures, diagrams, plots, drawings) used by the presenter to explain specific data or results. Avoid overloading the slides with too much information. Also, make sure the information on the slides follows a logical sequence that will keep the audience engaged.
  A practical rule for oral presentations is to use 3 minutes to discuss a slide. This limit allows the presenter to estimate the total number of slides able to be presented within the allocated period. A typical error observed across conferences and symposia is an overestimated

TABLE 6.4 Key Elements for a Successful Oral Presentation

Element	Description	Impact
Visual help	Digital slides. Digital photography.	Audience sees what is discussed. They can connect the information on the screen with the message from the oral discussion.
Color selection	Background, letters, frames.	Facilitates reading by the audience without distractions.
Letter's size and colors	Refer to **Figure 6.2** previously discussed.	Facilitates reading by the audience without distractions.
Amount of information per slide	Content is concise and brief. Pointing to the main idea.	Audience anticipates more details to be discussed by the presenter. Avoid reading the content of the slides.
Relevance of information	Highlights the most relevant information in the slides.	The main ideas are provided in the slides while the presenter expands on the topic.

number of slides and an underestimated required time for presentations. Make sure you practice and adjust your presentation so that it fits the assigned time.

King and Rupnow (1995) published an article providing specific guidance when preparing slides for oral presentations. The recommendations from King and Rupnow remain valid 30 years after the original publication. In summary, King and Rupnow recommend the following:

1. Visual Format
   The format of a slide is directly linked to the goal of the slide.
   - *Plots, Drawings, and Diagrams: A headline should be used summarizing the content of the plots, drawings, or diagrams presented in the slide. Additional information may be listed alongside in the form of bullets.*
   - *Plots (e.g., pie charts, bar charts, histograms) should be focused on a particular topic and avoid mixing plots presenting multiple topics.*
   - *All information must be clearly presented, concise, and self-explanatory.*

   Figure 6.4 presents an example of a slide with a plot.
2. Third Rule
   The slide is divided into three columns and three rows. The focal point is the square in the center of the slide. You will expand the

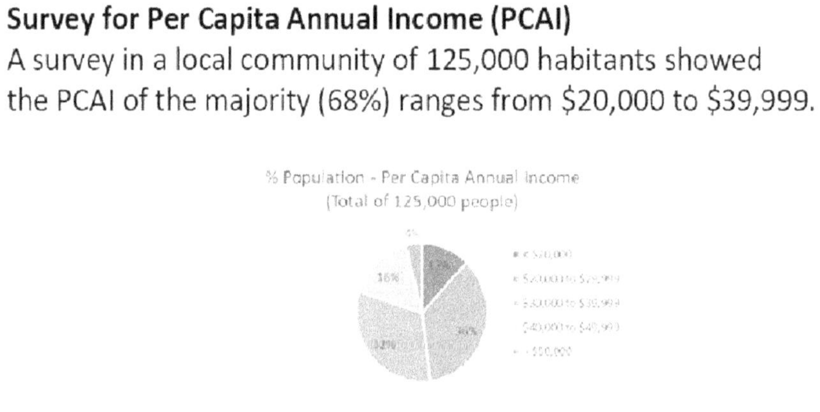

FIGURE 6.4 Example of a slide using a plot.

content to adjacent blocks (e.g., above, below, right, left) as needed. **Figure 6.5** presents an example of the rule.

3. Balance

   The content in slides can be either centered or aligned to either the left or right. **Figure 6.6** presents an example of both centered balance and right-side balance. The goal of the slide and the type of information presented should be the basis to decide where to place the information.

   - *Title of section and brief description – Centered or Aligned to the side.*

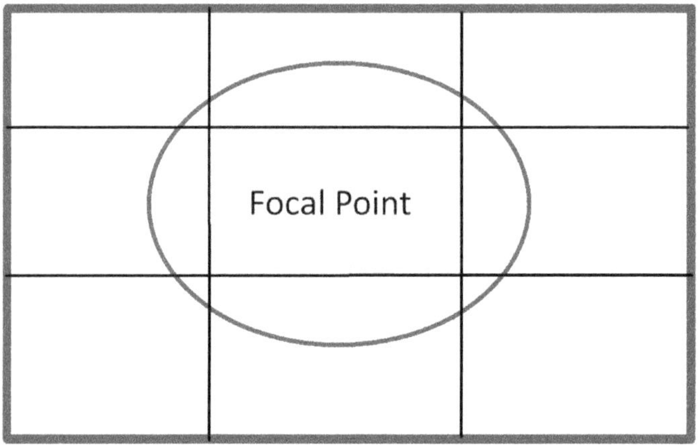

FIGURE 6.5 Example of the third rule.

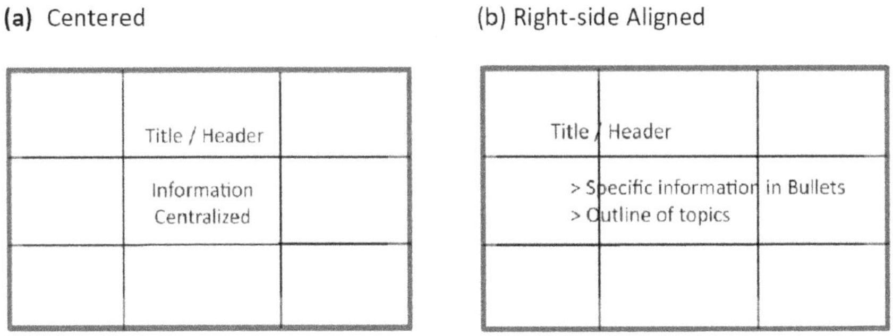

FIGURE 6.6 The concept of balance in slides: (a) centered and (b) right-side aligned. *Source*: Adapted from King and Rupnow, 1995.

- *Pictures, Plots, Drawings, Diagrams + Text – Centered.*
- *List of items and explanations – Aligned to the side.*

4. Gain and Maintain Attention
   The use of symbols or special characters to highlight and attract the attention of the audience is recommended. Symbols such as arrows, numbering, labels, logos, and changes in color are among the common options. All these options, among several more, are available in existing software used to create digital presentations.
5. Letters – Size and Fonts
   The size and style of the letters in a presentation are very important as the audience reads from the screen while the presenter speaks. The number of fonts available in software is significant, with few that are practical for presentations and a few that are not recommended. **Figure 6.7** shows example fonts that are commonly found in existing software (e.g., WORD®, PowerPoint®, Google Docs®, Apple Pages®) used to prepare presentations.
6. Colors
   Audiences are usually attracted to slides presenting a combination of colors vs. slides prepared only in black and white. However, it is important not to overuse colors. Too many colors in a slide may be too distracting from the technical content. The contrast of colors used as background and letters is critical to ensuring the readability of the text on a slide. In general, the use of dark backgrounds requires clear color letters, while light color backgrounds need darker letters. **Figure 6.8** shows some combinations of backgrounds and letters as examples of useful and non-useful combinations.

Font Type (Normal)	Size	Bold	Italics
Calibri	14	**Calibri**	*Calibri*
Arial	14	**Arial**	*Arial*
Courier	18	**Courier**	*Courier*
Time New Roman	12	**Time New Roman**	*Time New Roman*
Brush Script	11	**Brush Script**	*Brush Script*
Century Schoolbook	12	**Century Schoolbook**	*Century Schoolbook*
Aptos	12	**Aptos**	*Aptos*
Bahnschrift	12	**Bahnschrift**	*Bahnschrift*
Batang	14	**Batang**	*Batang*
Cambria	14	**Cambria**	*Cambria*
ENGRAVERS	11	**ENGRAVERS**	*ENGRAVERS*

FIGURE 6.7 Variety of fonts found in existing software.

Colors (Background/Letters)	Font Type (Normal)	Bold	Italics	Useful?
White/Black	Calibri	**Calibri**	*Calibri*	Yes
Yellow/Black	Calibri	**Calibri**	*Calibri*	Yes
Green/Black	Calibri	**Calibri**	*Calibri*	Yes
Yellow/Red	Calibri	**Calibri**	*Calibri*	Yes
Black/White	Calibri	**Calibri**	*Calibri*	Maybe
Black/Red	Calibri	**Calibri**	*Calibri*	Yes
White/Yellow				No
Black/Yellow	Calibri	**Calibri**	*Calibri*	Yes
Blue/Yellow	Calibri	**Calibri**	*Calibri*	Yes
Blue/Green	Calibri	**Calibri**	*Calibri*	No

FIGURE 6.8 Usefulness of colors (background and letters) on slides.

6.5 PUBLICATIONS

The publication of scientific work is one key goal for scientists. Scientists have multiple options for publication of research work (e.g., journals, magazines, books, news articles). What follows is some general guidance for the publication of scientific articles. It is the responsibility of the scientist to be familiar with specific requirements defined by the editorial boards.

Similar to reports and presentations, a publication is intended to consolidate the historical elements of research, experiments, test results, statistical analyses, and conclusions. A scientific article should contain the following sections (Hesselbeach et al., 2012; Schmidt et al., 2014; Harland, 2011):

- *Abstract*
  The abstract provides the readers with a summary of the topic under publication. The abstract may range from 250 to 500 words. In general, the abstract provides a high-level summary of the literature review, describes the goal of the study, summarizes the methodology, discusses key results, and concludes.
- *Introduction*
  This section is intended to provide the readers with a summary of the literature review, considering the key learnings obtained from already published scientific work. Also, the section should present the proposed hypotheses and the scope of the research.
- *Materials and Methods*
  Detailed descriptions of the methods and testing conditions are critical to make sure other scientists can repeat the study and achieve similar results. Experiments conducted using standardized methods published by recognized professional organizations such as the American Society of Mechanical Engineers (ASME), US Pharmacopeia (USP), American National Standards Institute (ANSI), and American Society for Testing and Materials (ASTM), and the Association of Official Analytical Chemists (AOAC), among others, will require the scientist to mention them and list the complete reference in the article.
  Pieces of equipment and parts required for the experiments should be described, including the manufacturer and operational conditions. If a piece of equipment is created or manufactured for the experiments, the scientist may consider publishing a high-level diagram or pictures. In this case, the scientist should consider applying for a patent to protect his or her invention.
- *Results and Discussion*
  Raw data from the experiments are carefully reviewed and analyzed. It is common to use tables and plots to summarize the results. Hypothesis testing uses statistical analyses to support the outcome of the experiments. Comprehensive and fact-based statements must be presented in the discussion of the results.
- *Conclusions*
  The conclusions are based on and must be supported by the results discussed in the previous section. The conclusion must clearly state the outcome of the hypothesis testing.

- *References*
  A complete list of the references compiled in the literature review should be presented in this section. The scientist must follow the recommended format of the journal.
- *Recognitions*
  This is an optional section. The author of the article or publication provides recognition to people or groups contributing to the work covered by the publication, or that sponsored the experimental work.

6.6 FINAL COMMENTS

The preparation of reports, presentations, and the publication of experimental work allows for communication to the scientific community. Every scientific work that rigorously follows the suggestions given here and reaches a level of statistical significance often results in some sort of publication and benefit for the scientific community. The amount of work associated with scientific research may be overwhelming; nevertheless, it is valuable work helping the human race. Examples of the contributions of scientific research are medical advances for the treatment and cure of diseases, new technologies helping workers in factories for the manufacture of better and more reliable products, new technologies for reforestation activities after wildfires, extending the shelf life of foods, and preventing foodborne illnesses.

Artificial intelligence (AI) tools such as ChatGPT®, Claude, Microsoft Copilot®, Google Gemini, and Notion AI are increasingly being used by researchers in drafting reports and presentations. To use these tools effectively, users must guide and refine them to match their writing style and the specific type of information needed. Like any technological resource, AI requires careful oversight, responsible data management, and thoughtful interpretation to ensure accurate and reliable outcomes. It is also important to recognize that AI platforms are designed with different priorities – some emphasize safety measures, others focus on speed, enterprise integration, or conversational style – so performance may vary depending on the tool selected.

6.7 REFERENCES

Harland, D.J. 2011. *STEM - Student Research Handbook: National Science Teachers Association*. NSTA Press, Arlington, VA.

Hesselbeach, R.A., Petering, D.H., Berg, C.A., Tomasiewicz, H., and Weber, D. 2012. A Guide to Writing a Scientific Paper: A Focus on High School Through Graduate Level Student Research. *Zebrafish* 9(4): 246–249. https://doi.org/10.1089/zeb.2012.0743.

King, J.W., and Rupnow, J. 1995. A Primer on Using Visuals in Technical Presentations. *Food Technology* 46(5): 157–170.

Schmidt, R.K., Smyth, M.M., and Kowalski, V.K. 2014. *Teaching the Scientific Literature Review: Collaborative Lessons for Guided Inquiry*. Bloomsbury Publishing, New York, NY.

Vega, H. 1980. Use of Glass as Fine Aggregate in Concrete Mixes. Jose de Diego High School Science Fair. Mayaguez, PR

Index